Hy-Q handbook of

Quartz Crystal
Devices

Hy-Q handbook of
Quartz Crystal Devices

DAVID SALT

Van Nostrand Reinhold (UK) Co. Ltd

© 1987 D. Salt

All rights reserved. No part of this
work covered by the copyright hereon
may be reproduced or used in any form
or by any means – graphic, electronic,
or mechanical, including photocopying,
recording, taping, or information
storage or retrieval systems – without
the written permission of the
publishers.

First published in 1987 by
Van Nostrand Reinhold (UK) Co. Ltd
Molly Millars Lane, Wokingham,
Berkshire, England

Typeset in 10/12pt Times by
Colset Private Limited, Singapore

Printed in Great Britain by
T.J. Press (Padstow) Ltd
Padstow, Cornwall

ISBN 0-442-31773-5

Contents

Foreword	ix
Preface	xi
Introduction	xiii
Historical survey	xiii
Piezoelectricity	xv
Elastic vibrations in solids	xviii
Quartz resonators	xix
Crystal oscillators and filters	xxi

Part 1 THE CRYSTAL AS A PHYSICAL DEVICE

1 Quartz resonators — 3

  1.1 Natural and cultured quartz — 3
  1.2 Crystallography and symmetry — 4
  1.3 Material constants — 5
  1.4 Modes of vibration — 8

2 Thickness mode resonators — 23

  2.1 Wave propagation in piezoelectric materials — 23
  2.2 Boundary conditions for thickness modes — 25
  2.3 Resonance frequencies and electromechanical coupling — 27
  2.4 Equivalent circuits — 30
  2.5 Temperature coefficients of frequency — 32
  2.6 Doubly rotated cuts — 35
  2.7 Rotated Y-cuts — 39
  2.8 AT-cut resonators — 41

3 Energy trapping — 45

  3.1 Wave propagation in thin plates — 45
  3.2 Resonance frequencies for finite plates — 49
  3.3 Coupled modes in finite resonators — 51
  3.4 Energy trapping — 55
  3.5 Mass loading — 58
  3.6 Contouring and bevelling — 59

Part 2 MANUFACTURING TECHNIQUES

4 Optical processing — 63

  4.1 X-ray orientation — 64

	4.2 Sawing, dicing and rounding	69
	4.3 Lapping and polishing	70
	4.4 Contouring and bevelling	74

5 Electrical processing — 77

	5.1 Cleaning and etching	77
	5.2 Base-plating	78
	5.3 Mounting and bonding	79
	5.4 Adjusting to frequency	81
	5.5 Sealing	85
	5.6 Photolithographic techniques	86

Part 3 THE CRYSTAL AS A CIRCUIT ELEMENT

6 Equivalent circuit analysis — 91

	6.1 Equivalent circuits	91
	6.2 Lossless circuit analysis	92
	6.3 Lossy circuit analysis	95
	6.4 Effective crystal parameters with a series load capacitor	103

7 Characteristics of AT-cut crystal resonators — 107

	7.1 Blank diameter and geometry	107
	7.2 Motional capacitance and shunt capacitance	110
	7.3 Q factor and time constant	112
	7.4 Frequency-temperature characteristics	114
	7.5 Unwanted responses	117
	7.6 Ageing and short-term frequency stability	117
	7.7 Non-linear effects	119

8 Measurements — 120

	8.1 Introduction	120
	8.2 CI meters	120
	8.3 Zero phase measurement systems	121
	8.4 Automated measurement systems	123

Part 4 CRYSTAL OSCILLATORS AND FILTERS

9 Quartz crystal oscillators — 127

	9.1 Introduction	127
	9.2 Elementary circuits	128
	9.3 Frequency stability: circuit considerations	131
	9.4 Frequency stability: crystal characteristics	134
	9.5 Types of crystal oscillator	136

10 Quartz crystal filters — 141

	10.1 Introduction	141

10.2 Elementary circuits: discrete crystal filters	144
10.3 Elementary circuits: monolithic crystal filters	151
10.4 Limitations on crystal filter performance	156

Appendices 160

1 Explanation of piezoelectric effect	160
2 Vectors and tensors	162
3 Continuum mechanics	172
4 Linear piezoelectric theory	190
5 Coordinate transformations and crystal symmetries	197
6 Wave propagation in isotropic plates	208

References 220

Index 225

Foreword

The applications for quartz crystals have grown from the early days in frequency standards and in amateur radio communications to the present wide spectrum of applications.

This Handbook will enable you to understand the importance of quartz crystals for frequency control, selectivity and other applications using quartz.

These devices, the heart of communication systems, are a real necessity in our fast moving time. This book on quartz crystal devices will give you an understanding of the electrical–mechanical function of quartz with its limitations, along with fabrication techniques that will cover precision X-ray, grinding, polishing, vapor deposition, mounting techniques, encapsulation, and measurement systems.

Dr Salt provides information in the Handbook for guidance in specifying quartz devices to meet the application best suited for the design engineer. Since quartz crystals at present cover a wide range of frequency control from a few parts in 10^9 ageing and as much as 10^{14} short-term stabilities, it is important that design engineers understand the complexities of manufacture and the limitations of quartz. This will allow the engineer to specify more intelligently for his application, performance, and economic advantage.

Dr Salt's book will be most welcome to design engineers.

L. W. McCoy
President Emeritus
McCoy Electronics Company
Mt. Holly Springs, PA, USA

Preface

This book is intended to be useful to engineers concerned with the design of systems that involve frequency control or selection, or, in short, *frequency management*. Bulk wave quartz crystal resonators and associated devices are key components of such systems and yet basic information about quartz crystal products is not easily available.

There are three reasons for this situation. First, the principles of operation of crystal devices do not generally form part of the curriculum of electronic engineering courses. Second, the major texts in the field, those of Cady, Mason, Vigoreux and Booth, and Heising, are either out of print or not readily available, and in any case are generally aimed at the specialist in the field rather than the general user. Third, many of the more modern developments in the crystal field are scattered throughout a wide variety of learned journals, symposia and conference proceedings, and technical reports, and are quite simply difficult to find.

The recent books by Bottom (1982), Frerking (1978), Matthys (1983), and Parzen (1983) have all contributed towards closing this 'information gap', while the recent two-volume work edited by Gerber and Ballato (Gerber and Ballato, 1985) covers the whole field of precision frequency control including devices such as atomic and laser frequency standards on a 'state-of-the-art' level. The emphasis of the present work is very much tutorial, the main objective being to provide enough background material on both the design and manufacture of quartz devices to give an understanding of the practical and theoretical limitations on the performance of commercially available products. In some cases, theoretical considerations predominate but in many instances the practical limitations due to the complexities of the manufacturing process are the determining factors. Hence some fairly detailed descriptions of manufacturing processes are included.

Following an Introduction giving a broad view of the field, the book is organized in four main parts. Part 1 deals with the crystal resonator as a physical device. A survey of the wide range of different types of resonator is followed by a more detailed review of thickness mode resonators with particular emphasis on the AT-cut crystal which is the most widely used in modern practice. Part 1 provides the physical basis for understanding the operation of a resonator in terms of the normal modes of vibration of a piezo-electric solid. To avoid obscuring the main points of the text, the more theoretical aspects of the topics covered are restricted to a number of appendices. Part 2 moves on to discuss the manufacture of crystal resonators, with the emphasis on the AT type. Low frequency crystal cuts are only briefly discussed in connection with modern photolithographic manufacturing methods.

Part 3 considers the crystal as a circuit component in terms of its equivalent circuit. The elements of the equivalent circuit are related to the physical characteristics of the device discussed in Part 1. The resonant frequencies and impedance and admittance characteristics of the crystal as represented by its equivalent circuit are discussed in detail, and the available measurement techniques are reviewed.

In Part 4 the key characteristics of bulk wave crystal oscillators and crystal filters are discussed. After a description of elementary oscillator circuits of the Pierce type, the main factors affecting the frequency stability of crystal oscillators are reviewed. This is followed by a survey of the main categories of packaged oscillator currently available. In the final chapter, the basic circuit configurations of narrow, intermediate and wideband crystal filters are explained, followed by a discussion of the practical limitations on crystal filter performance.

The author wishes to thank the management and staff of Hy-Q International for providing both the opportunity to undertake the task of writing this book, and also for their continuing support and encouragement throughout its duration.

David Salt

Introduction

HISTORICAL SURVEY

The early history of the subject is covered in detail in Cady's classic work *Piezoelectricity* (Cady, 1964), and in Heising (1946). The direct piezoelectric effect, the production of electricity by the application of pressure, was discovered by the brothers J. and P. Curie in 1880. The converse effect was predicted by Lippmann in 1881 and confirmed by the Curies in the same year. The first major application of piezoelectricity was in Langevin's work on submarine detection using quartz transducers to generate and detect underwater acoustic waves, started in 1917. Nicolson in 1919 described several devices, such as loudspeakers, microphones and sound pick-ups based on the piezoelectric properties of Rochelle salt, and this was followed by Cady's publication in 1921 of papers describing the first quartz crystal controlled oscillator. From this time considerable research effort was devoted to the field of quartz crystals.

In 1926, the station WEAF in New York became the first broadcast transmitter to use quartz crystal control. At first the further application of crystal control in radio communications was hampered by the temperature dependence of the crystal frequency, which made necessary the use of cumbersome temperature control equipment, especially in applications such as aircraft communications. This obstacle was effectively removed by the introduction in 1934 of crystal cuts with very low temperature coefficients of frequency (Lack *et al.*, 1934). The early work on frequency control was paralleled by work on frequency selective devices, and the quartz crystal filter had become an essential component of telephone line transmission systems by 1939.

With the approach of World War II, it was recognized by the military authorities that crystal control was essential if the full benefit of radio communications was to be obtained in battle. The consequent demand for quartz crystals resulted in a tremendous expansion of the then embryo crystal industry. In the United Kingdom, production increased from 10 000 units in 1938 to 1.5 million units in 1945 (Vigoreux and Booth, 1950, p. 1), and in the USA the Western Electric company's crystal production increased 750-fold during the war period (Heising, 1946, p. 8). This expansion was only possible thanks to the concerted efforts of many people in solving the many design and manufacturing problems that were encountered, and the subsequent dissemination of this knowledge in such works as those of Heising and of Vigoreux and Booth, helped to lay the foundation for the subsequent growth of the industry.

The post-war period has seen sustained development in the traditional application areas of frequency control and selection in radio and line communication systems. Shortages of natural quartz during the war years inspired both a search for alternative piezoelectric materials and also research into techniques for producing cultured quartz crystals. The success of the latter program is evident insofar as cultured quartz, commercially available from several sources, has now practically replaced natural quartz as the industry's raw material. Improvements in design methods and manufacturing techniques have produced successive advances in crystal unit performance, particularly in regard to miniaturization, increased frequency stability, and increased frequency range. Crystal filters, previously mainly restricted to line transmission systems, have been increasingly used since the late 1950s in other applications, particularly as channel filters in mobile radio systems. This trend was accelerated by the introduction of the monolithic crystal filter in the middle 1960s.

Particularly in the case of radio communications, the historical demand for quartz crystals has for the most part been characterized by requirements for many different frequencies, with relatively small quantities on any particular frequency. The typical end use was that of a multi-channel radio set requiring two crystals per channel, one for transmit, one for receive. Generally, different users operated on different channel frequencies, and therefore required 'channel' crystals unique to themselves. Consequently the crystal industry developed with a manufacturing philosophy oriented to small batch production in an environment where customer service, in terms of rapid response to a customer's specific demands, was paramount.

In recent years this orientation has proved increasingly inappropriate. The impact of frequency synthesis techniques has reduced the overall requirement for channel crystals, and substituted a growing demand for stable reference frequency sources that are in principle better suited for standardization. Coupled with the demands for greater frequency stability resulting from the use of narrow channel spacings at higher and higher channel frequencies, the modern trend is towards the use of packaged temperature-compensated crystal oscillators in synthesized systems. Just as 'de facto' industry standards have developed for mobile radio channel filters on frequencies of 10.7 and 21.4 MHz, so it is to be expected that a similar standardization of packaged oscillator specifications will come about. Emphasis in manufacturing will then have to be placed on achieving high quality and reliability at low cost, with the premiums presently available for rapid delivery of custom products such as channel crystals an anachronism.

These considerations are even more appropriate to the many new applications for quartz crystals that have developed in recent years. Crystals for colour television, for clocks and watches, for video tape recorders and for all kinds of timing applications such as microprocessor clocks, are all required in high volumes and on fixed frequencies. So too are the simple packaged oscillators generically known as 'clock' oscillators that are tending to take the place of discrete crystals in the timing applications just mentioned. These products are essentially mass-produced commodity items, representing a major, fundamental, change from the traditional crystal products business. A further feature of the transformation of the crystal market brought about by these new commodity products is that in pursuing the objective of miniaturization, specifically in the case of crystals for electronic watches, photolithographic manufacturing techniques have been introduced that have in turn spawned a whole range of new miniature crystal units with applications far beyond that originally envisaged for watches.

In summary, in the century or so since the discovery of piezoelectricity, the history of quartz crystal devices can be divided roughly into five periods. In the first period, from 1880 to about 1920, piezoelectricity remained a scientific curiosity. The second period from 1920 to 1939 can be described as the classical period, when the foundations were being laid for the forced growth of the third period, World War II. The 20 years immediately following the war, up to about 1965, constituted a fourth period of steady growth and orderly development, and the fifth period, from 1965 to the present has seen a succession of fundamental changes in both markets and techniques which can fairly justify the epithet 'revolutionary'.

PIEZOELECTRICITY

Piezoelectricity is literally 'pressure electricity', the prefix *piezo* being derived from the Greek 'to press'. The direct piezoelectric effect discovered in 1880 by the brothers Curie refers to the electric polarization of certain materials brought about by the application of a mechanical stress. The converse effect, predicted by Lippman and confirmed by the Curies in 1881, refers to the deformation produced in the same materials by the application of an electric field. Details of these early researches are given by Cady (1964).

All dielectric materials exhibit the phenomenon of *electrostriction*, that is they are deformed when placed in an electric field. Piezoelectric materials are generally distinguished from purely electrostrictive ones by the relatively much larger piezoelectric deformation and by the fact that the piezoelectric deformation is reversible. Thus if, for example, a rod of quartz is cut in such a way that an applied field causes an elongation of the rod, reversing the direction of the field will cause the rod to contract, whereas in a non-piezoelectric material, whatever deformation is caused will be independent of the direction of the field. Quantitatively, a piezoelectric deformation is essentially linearly dependent on the applied field, whereas an electrostrictive deformation depends on the square of the field. With the exception of ferroelectric materials and some piezoelectric ceramics, the electrostrictive effect is generally orders of magnitude smaller than the converse piezoelectric effect where the latter exists. This is certainly true of quartz.

The reversible nature of the piezoelectric effect implies that piezoelectric materials must be anisotropic, that is that their physical properties must depend on direction within the material. More precisely, such materials cannot have a centre of symmetry, for with a centre of symmetry, the reversal of an applied field could have no significance relative to the material's internal structure. An explanation of the origin of the piezoelectric effect in terms of an assumed molecular structure for quartz was first given by Lord Kelvin soon after the Curies' initial discovery, and although his proposed structure has since been proved incorrect by X-ray crystallography, his explanation can still be accepted as a qualitative and heuristic guide to understanding the phenomenon. Kelvin's explanation (Appendix 1) hinges on the assumption of distributions of charges on the molecular level which are such that certain deformations of the structure bring about a separation of the centres of gravity of the positive and negative charges and therefore the production of an electric dipole moment or electrical polarization.

Physically, the polarization in a piezoelectric material can be regarded as having two components, one directly related to the mechanical deformation of the material by some such mechanism as discussed above, and the second related to the electric field in the material by the normal mechanisms of dielectric polarization. These two components can in principle be isolated in suitable experiments. For example, if a sample of material was clamped in such a way that no mechanical deformation was possible, then the polarization of the sample would be entirely due to the applied electric field. Similarly, if the surface of the sample was coated with a conducting layer maintained at a constant electric potential so that the internal electric field was constrained to be zero, then the polarization of the sample would be entirely due to its deformation. In the general case, the polarization would be the sum of these two components, and be written in the form

$$P = kE + eS$$

where P, E and S are the polarization, electric field and mechanical strain, respectively, and k, e are, respectively, the susceptibility and the piezoelectric stress constant.

The deformation or strain in a piezoelectric material can likewise be thought of as having two components, the piezoelectric strain due to the electric field and the additional strain due to any mechanical stresses in the material. Again, these two components can in principle be separated in ideal experiments. Thus the piezoelectric strain would be that observed in the material in the presence of an electric field and complete absence of any mechanical forces, whether body forces such as that due to gravity, or surface tractions. The purely elastic strain would be that due to applied stresses in a sample of material where the field was constrained to be zero. The general expression for the strain would then be written

$$S = sT + dE$$

where T is the stress in the material and s, d are, respectively, the elastic compliance and the piezoelectric strain constant.

It is tempting, but potentially confusing, to regard the piezoelectric strain as being the result of a piezoelectric stress due to the electric field. By definition, a stress is a force per unit area of a surface exerted by the material on one side of the surface and acting on the material on the other side. The piezoelectric strain is, however, caused by the electric field acting at long range and not by transmission of a stress through the body of the material. The distinction between the states of stress in an elastically deformed body as compared to a body similarly deformed by an electric field through the agency of the piezoelectric effect can be illustrated by considering the case of a long thin rod. Suppose first that such a rod is stretched by the application of equal and oppositve longitudinal forces at its ends. The rod as a whole is in equilibrium and in a state of tensile stress, but if the rod is cut in two the equilibrium is destroyed. The two halves of the rod will revert to a state of zero strain and at the same time be accelerated by the unbalanced external forces acting on them. The situation is quite different in the case of the same rod stretched by the same amount by an electric field through the piezoelectric effect. Cutting the rod in this case now produces no effect whatsoever. The two halves remain in the same state of strain and remain stationary, thus indicating that there were no forces acting across the cross-section of the rod where the cut was made, that is, there is zero stress in the material.

Because of the tensor nature of the stress T and strain S and the necessarily anisotropic character of the material, it is not possible to interpret the material constants k, e, s and d as scalar quantities. P and E are vector quantities specified by their three components with respect to some coordinate system, so that the susceptibility k must be treated as a second rank tensor quantity. The stresses and strains T and S are themselves symmetric second rank tensors and so the piezoelectric constants e and d relating the vector quantities E and P to T and S must be treated as third rank tensors, and the elastic constants c and s are correspondingly fourth rank tensors. A brief discussion of tensor formalism and the formulation of the linear theory of piezoelectricity is given in the appendices.

ELASTIC VIBRATIONS IN SOLIDS

If a sample of elastic material is deformed by the application of external forces and those forces are suddenly removed, then vibrations will be set up in the material, gradually dying away as the energy stored in the material by the work done in the initial deformation is dissipated by the various internal and external loss mechanisms present. This phenomenon is familiar in such well known instances as bells, gongs, tuning forks and xylophones. In musical instruments, the pleasant effect of the resulting sound is due to the fact that the structure or geometry of the device results in the generation of a relatively small number of harmonically related frequencies. Historically, the design of such instruments as bells would have been arrived at empirically, and in general the problem of calculating the frequencies of a vibrating body in terms of its dimensions and material properties is very difficult. Nevertheless, the solution of the problem in the case of quartz crystal resonators is a necessary part of their design.

There are two situations to be considered. The first is the case of the free vibrations of a body set in motion by arbitrary forces and then released; the second the case of forced vibrations, where an external periodic driving force is present throughout the motion. In the former case, the vibrations vary from instance to instance depending upon the initial conditions; this is illustrated by the variations in tone that result from striking a bell in different positions on its surface, and can be explained physically in terms of the concept of normal modes of vibration. In a normal mode, all points of the body share in a (damped) simple harmonic motion and have the same phase and frequency. Thus when any one point goes through its equilibrium position, so too do all the other points, and when the same point reaches its extreme position, so too do the others. Analysis shows that there are generally an infinite number of such modes, and that an arbitrary free motion of the body can be regarded as a superposition of normal modes with different amplitudes and phases determined by the initial conditions. Practically speaking, it is impossible to excite a single normal mode, since this would require the initial deformation precisely to mirror that of the particular mode in question throughout the material. A good approximation to a single mode resonator can however be obtained by the proper choice of geometry and dimensions.

In the presence of an harmonic driving term of a given frequency, the onset of vibrations will be accompanied by the excitation of normal modes. In a real, lossy, material these form transients dying away exponentially, leaving a steady-state vibration of the same frequency as the driving force. The amplitude of the forced vibrations will generally be of the same order of magnitude as the static deformation produced by a force equal to the amplitude of the periodic driving force, except when the driving frequency is close to the frequency of a normal mode of the device, when resonance occurs. At resonance, the amplitude of vibration increases by an amount limited only by dissipation, in analogy to the increase in current observed when the frequency of an applied voltage is swept through the resonance frequency of a series tuned LCR circuit. The analogy is almost exact for frequencies close to the resonance frequency, but as the electrical circuit has only the one resonance frequency, it does not reflect the multiplicity of resonance frequencies that exist in the mechanical vibrator.

In resonant systems where the resonator material has low losses and the driving force is only loosely coupled to the actual resonator, the driving term can be largely ignored in respect of the calculation of resonance frequencies, simply being regarded as a means of maintaining the vibration. Both these conditions are true in the case of quartz resonators, thus simplifying the analysis of the device by allowing attention to be restricted to free vibrations only.

QUARTZ RESONATORS

A piezoelectric resonator consists of a piece of piezoelectric material precisely dimensioned and oriented with respect to the crystallographic axes of the material and equipped with one or more pairs of conducting electrodes. By means of the piezoelectric effect an electric field applied between the electrodes excites the resonator into mechanical vibration. The amplitude of vibration is negligibly small except when the frequency of the driving field is in the vicinity of one of the resonator's normal modes of vibration and resonance occurs. Near resonance, the amplitude of vibration increases and, again by virtue of the piezoelectric effect, the electrical impedance of the device changes rapidly.

Many different substances have been investigated as possible resonator materials, but for many years quartz resonators have been preferred in satisfying needs for precise frequency control and selection. Compared to other resonators, for example LC circuits, mechanical resonators such as tuning forks, and piezoelectric resonators based on ceramics or other single-crystal materials, the quartz resonator has a unique combination of properties. The material properties of single-crystal quartz are both extremely stable and highly repeatable from one specimen to another. The acoustic loss or internal friction of quartz is particularly low, leading directly to one of the key properties of a quartz resonator, its extremely high Q factor. The intrinsic Q of quartz is 10^7 at 1 MHz. Mounted resonators typically have Q factors ranging from tens of thousands to several hundred thousand, orders of magnitude better than the best LC circuits.

The second key property of the quartz resonator is its stability with respect to temperature variation. Depending on the shape and orientation of the crystal blank, many different modes of vibration can be used and it is possible to control the frequency–temperature characteristics of the resonator to within close limits by an appropriate choice. The most commonly used type of resonator is the AT-cut where the quartz blank is in the form of a thin plate cut at an angle of about 35° 15′ to the optic axis of the crystal. The AT-cut has a frequency–temperature coefficient described by a cubic function of temperature, which can be precisely controlled by small variations in the angle of cut. This cubic behaviour is in contrast to most other crystal cuts which show parabolic temperature characteristics, and makes the AT-cut well suited to applications requiring a high degree of frequency stability over wide temperature ranges.

The third essential characteristic of the quartz resonator is related to the stability of its mechanical properties. Short and long term stabilities manifested in frequency drifts of only a few parts per million per year are readily available even from commercial units. Precision crystal units manufactured under closely controlled conditions are second only to atomic clocks in the frequency stability and precision achieved.

The AT-cut resonator uses the thickness shear mode of vibration. A standing wave is set up in the crystal blank by the reflection at both major surfaces of transverse waves travelling in the thickness direction. The major mechanical displacement is in the plane of the crystal, at right angles to the direction of wave propagation. At resonance, an odd number of half wavelengths is contained in the thickness of the plate, the thickness therefore being the primary frequency determining dimension. The AT-cut is thus an example of a thickness mode resonator. AT-cuts are commonly manufactured in the frequency range from about 1 MHz to 200 MHz and above, and in this range are usually the optimum choice for most applications. The AT-cut is, however, sensitive to stresses in the body of the resonator, whether caused by temperature gradients due to rapid external temperature changes, or by external forces. For applications where extreme stability is required, or where severe environmental conditions are likely to be encountered, this stress sensitivity is a drawback, and newer crystal cuts such as the SC have been developed that minimize these effects. The SC-cut is also a thickness mode resonator, and as such is basically available in much the same frequency range as the AT, though its commercial availability is presently much more limited.

Below about 1 MHz, thickness-mode resonators generally become too cumbersome and unwieldy for general use, and other modes of vibration are used. At very low frequencies, below about 100 kHz, flexural mode and length extensional devices are used. In both cases the crystal is cut in the form of a long thin bar with the length of the bar being the primary frequency determining dimension. Above 100 kHz face shear devices, such as the CT- and DT-cuts, are used, where the crystal element is a square or rectangular plate, and the length of the edges determines the frequency. The GT-cut also uses a rectangular plate vibrating in coupled length and width extensional modes, and is unique among crystal cuts in having an almost zero frequency–temperature coefficient over a wide range of temperature. The traditional method of manufacture for all these low frequency crystal cuts involved mounting the crystal element by means of thin wires attached to the crystal at nodal points and soldered or welded to the crystal holder. Such wire-mounted units are intrinsically less robust than the higher frequency thickness mode resonators. Recent developments triggered by the huge demand for miniature crystal units for electronic watches have however initiated a renaissance in the technology of low frequency crystal manufacture, typified by the introduction of a miniature GT-cut crystal in which both resonator and support structure are fabricated as one piece using photo-lithographic techniques. These techniques, originally developed for the mass production of watch crystals of the tuning fork type using flexural vibrations, are now also being applied to the miniaturization of AT type resonators.

CRYSTAL OSCILLATORS AND FILTERS

Crystal filters and oscillators are devices in which one or more resonant circuits are realized by crystal resonators in order to benefit from their superior Q, and their frequency stability with respect to temperature and time. In the case of filters, the high Q available from quartz resonators in particular is essential in the practical realization of effective narrowband filters, and crystal filters have long been a key component in frequency division multiplex (FDM) line transmission systems. In modern radio communications practice, crystal filters play an equally essential role in determining the selectivity of radio systems. The latter role has largely developed since the 1950s and was greatly accelerated by the introduction of the monolithic crystal filter in the 1960s. In the monolithic filter, not just one but two or more coupled resonant circuits are realized on a single piece of quartz, resulting in large cost and space savings and much increasing the attractiveness of crystal filters to the equipment designer.

In the early years following Cady's pioneering work on the crystal controlled oscillator, the only applications envisaged were in frequency standards and other purely scientific areas. The need for more general crystal control of oscillators was not recognized until 1926, when the problems of poor quality reception being experienced by listeners to the radio station WEAF in New York were solved by the introduction of crystal control of the transmitted signal (Heising, 1946). Subsequent developments in both communications and crystal technology have brought about a situation where crystal control is vital to the present intensive use of the radio frequency spectrum.

For much of the period from Cady's work to the present, the design of the crystal oscillator was the province of the user, the equipment manufacturer. The crystal manufacturer supplied only the crystal. Although there have always been exceptions to this general rule, the growth of a volume market for packaged crystal oscillators is a relatively recent phenomenon, driven by several factors including the impact of frequency synthesis techniques, the increasing need for temperature compensated devices, and latterly the scarcity of design engineers familiar with analog circuits. In non-synthesized multi-channel radio equipments, the normal design approach is to obtain the desired channel frequencies by switching appropriate crystals, and there is no scope for packaged oscillators. However, in a synthesized equipment only one reference oscillator is required from which all necessary channel frequencies are derived, and it is clearly advantageous for the equipment designer to be able to specify the reference oscillator as a bought-in component rather than have to go through a complete design exercise. This is especially true when the stability requirements are such that temperature compensation of the oscillators is required. It is a feature of the temperature compensation problem that whatever method of compensation is adopted has to be tailored to individual crystals. This means that in a non-synthesized system all channel crystals have to be separately compensated, whereas in a synthesized system only the reference frequency needs compensation. Again from the equipment designer's viewpoint, it is surely preferable to be able to specify the stability required in a packaged oscillator and not to have to worry about the niceties of temperature compensation.

The emphasis in modern electronics is very much on digital techniques, which are increasingly being used in areas such as communications previously dominated by analog methods. For an engineer with a digital electronics background, the design of a crystal controlled oscillator is likely to present problems that can be avoided by the specification and procurement of a packaged oscillator. This is particularly relevant in the case of wholly digital systems where an oscillator is required simply to produce timing signals to allow the synchronization of different digital processes.

As a result of these various factors a wide range of packaged crystal oscillators is now commercially available. At one extreme are the 'clock' oscillators designed for just such timing applications as mentioned above. These are manufactured in large volumes in standard dual-in-line configurations. Their frequency stability is relatively poor, but adequate for the application, and they are of course designed to interface directly with digital circuits. At the other extreme are crystal oscillators incorporating miniature crystal ovens and achieving frequency stabilities of parts in 10^8 over wide temperature ranges. In between these extremes lies a large variety of different oscillator types, including, for example, voltage controlled oscillators

(VCXOs) and temperature compensated crystal oscillators (TCXOs), which allows an equipment engineer to concentrate on the larger issues of system design rather than the details of crystal oscillator design.

In crystal filters, too, there is currently available a wide range of devices for performing narrowband filtering functions. Apart from line transmission applications, the bulk of filters sold are IF or 'channel' filters for VHF and UHF radio systems. Other filter types include SSB filters for eliminating unwanted sidebands, 'roofing' filters for double-conversion HF receivers, narrowband bandstop filters, and various filters with specialized phase and group delay responses. In the channel filter sector there is a good deal of standardization, but in other sectors the crystal filter market is still primarily one for custom-built devices.

Part 1
The crystal as a physical device

1 Quartz resonators

1.1 NATURAL AND CULTURED QUARTZ

Quartz is a crystalline form of silicon dioxide, SiO_2. It is a hard, brittle, transparent material with a density of 2649 kg m^{-3} and a melting point of 1750°C. When quartz is heated to 573°C its crystalline form changes. The stable form above this transition or inversion temperature is known as 'high-quartz' or 'beta-quartz', the stable form below 573° is known as 'low-quartz' or 'alpha-quartz'. For resonator applications, only alpha-quartz is of interest and unless stated otherwise the term 'quartz' in the sequel always refers to alpha-quartz. Quartz is insoluble in ordinary acids, but soluble in hydrofluoric acid and in hot alkalis.

Quartz is one of the commonest naturally occurring crystalline minerals, sand for example being largely made up of grains of quartz produced by the weathering of larger crystals. Despite this natural abundance, crystals of sufficient size and purity for processing into quartz resonators are very rare. The occurrence and characteristics of natural quartz are fully described by Willard in Chapter IV of Heising (1946). For all but very exceptional requirements, natural quartz has now been superseded by cultured quartz in resonator manufacture. Cultured quartz has a long history (Cady, 1964), but no significant progress was made until the shortage of raw material for the war effort in 1939–45 spurred an intensive post-war program of research on both sides of the Atlantic. Cultured quartz is now routinely grown from aqueous alkaline solutions under conditions of high pressure and temperature in massive underground steel autoclaves. The lower part of the autoclave is maintained at a temperature of about 400°C and contains nutrient in the form of pure silica (SiO_2). At this temperature and at pressures of the order of a thousand atmospheres, the solubility of silica is relatively high and a saturated solution is formed. Convection currents transport the saturated solution up to the upper part of the autoclave which is maintained at a slightly lower temperature of about 350°C. At this lower temperature the solution is supersaturated and quartz is deposited on seed crystals suspended in the cooler region of the autoclave. Over periods of many days or weeks crystals of substantial size can be grown. Brice (1985) gives more details of the process.

1.2 CRYSTALLOGRAPHY AND SYMMETRY

Alpha-quartz belongs to the crystallographic class 32. It has a single axis of threefold symmetry (trigonal axis) and perpendicular to this three axes of twofold symmetry (digonal axes). The digonal axes are spaced 120° apart and are polar axes, that is a definite sense can be assigned to them. The presence of polar axes implies the lack of a centre of symmetry and is a necessary condition for the existence of the piezoelectric effect. The digonal axes are known as the electric axes of quartz, since in the original experiments of the Curies the electrical polarization produced by pressure was developed along a polar axis. In crystals with fully developed natural faces the two ends of each polar axis can be differentiated by the presence or absence of the s and x faces (Fig. 1.1). When pressure is applied in the direction of the electric axis a negative charge is developed at that end of the axis modified by these faces. The trigonal axis, also known as the optic axis or c-axis, is not polar, since the presence of digonal axes normal to it implies that the two ends of the trigonal axis are equivalent. Thus no piezoelectric polarization can be produced along the optic axis.

Alpha-quartz is an optically active material. When a beam of plane-polarized light is transmitted along the optic axis a rotation of the plane of polarization occurs, the amount of the rotation depending on the distance traversed in the material. The sense of the rotation can be used to differentiate between the two naturally occurring forms of alpha-quartz known as left quartz and right quartz. In right quartz the plane of polarization rotates clockwise when seen by an observer looking towards the source of light, and in left quartz it rotates anti-clockwise. In crystals with fully developed s and x faces their orientation with respect to the major prism faces also provides a means of differentiation between right and left quartz (Fig. 1.1).

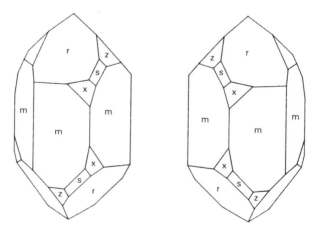

Fig. 1.1 Right and left hand quartz crystals.

The handedness of cultured quartz depends on the handedness of the seed crystals used. Most cultured quartz produced is right quartz, whereas in natural quartz left and right forms are about equally distributed. Either form can equally well be used in the manufacture of resonators, but material in which left and right forms are mixed, that is *optically twinned* material, cannot be used. *Electrically twinned* material is all of the same hand, but contains regions where the sense of the electric axis is reversed, thus reducing the overall piezoelectric effect. Such material is also not suitable for resonator application. The presence of twinning and other defects in natural quartz crystals is the major reason for the shortage of suitable natural material, and the absence of significant twinning in cultured quartz constitutes one of its main advantages.

When alpha-quartz is heated to above 573°C, the crystalline form changes to that of beta-quartz, which has hexagonal rather than trigonal symmetry. On cooling down through 573° the material reverts to alpha-quartz, but in general will be found to be electrically twinned. Similarly the application of large thermal or mechanical stresses can induce twinning, so it is necessary in resonator processing to avoid any such thermal or mechanical shocks.

The precise description of the physical properties of anisotropic materials requires the definition of a coordinate system. In the past several different conventions and standards have been used; the coordinate system used here follows the 1978 *IEEE Standard on Piezoelectricity* (IEEE, 1978). The reference system is a rectangular system of coordinates $OXYZ$ in which the $+Z$ direction is chosen parallel to the optic axis. The X direction is chosen to be parallel to an electric axis with the positive sense in right quartz being towards the end of the axis modified by the s and x faces. The Y direction is then chosen so as to form a right-handed coordinate system. In left quartz the sense of the X direction is reversed, so that the positive sense is away from the modified end of the electric axis, but still the Y direction is selected to keep a right-handed system. Adoption of this standard results in all the piezoelectric constants of left quartz having the opposite sign to those of right quartz.

1.3 MATERIAL CONSTANTS

The behaviour of quartz resonators, at least in the linear theory, is described by the field equations and boundary conditions discussed in Appendix 4. In order to be able to apply these equations a knowledge of the material constants appearing in them is required. The relevant parameters are the elastic constants c_{ikmp}, the piezoelectric constants e_{ikm}, the dielectric constants ϵ_{ik} and the density ρ. In addition, in order to be able to understand the variation of resonator characteristics with temperature, knowledge is required both of the thermal expansion coefficients and also of the temperature coefficients of the constants already listed.

In a completely anisotropic material, there are 21 independent elastic constants, 18 piezoelectric constants and 6 dielectric constants. Along with the density, there are therefore a total of 46 independent quantities needed to describe the relevant properties of the material at a given temperature. In the context of resonator theory, it is usual to describe the temperature dependence of a quantity $f(T)$ by giving the temperature coefficients $T_f^{(r)}$ of order r for $r = 1$, 2 and 3. The $T_f^{(r)}$ are defined for a given reference temperature T_0 by expanding $f(T)$ in a Taylor series about T_0 and setting

$$T_f^{(r)} = f^{(r)}(T_0)/(r! f(T_0))$$

where $f^{(r)}(T_0)$ denotes the rth derivative of f with respect to its argument T, evaluated at T_0.

Then neglecting terms higher than third order in $(T - T_0)$, $f(T)$ can be written

$$f(T) = f(T_0)\{1 + T_f^{(1)}(T - T_0) + T_f^{(2)}(T - T_0)^2 + T_f^{(3)}(T - T_0)^3\}$$

In a completely anisotropic material, the number of temperature coefficients required is therefore $3 \times 46 = 138$.

The thermal expansion coefficients of order r are defined in a similar way by expanding the thermally induced strain in a Taylor series and retaining up to third order terms. If S_{ik} is the strain at temperature T referred to the configuration of the material at temperature T_0 then the expansion coefficients of order r are the coefficients of $(T - T_0)^r$ in the expansion

$$S_{ik}(T) = \alpha_{ik}^{(1)}(T - T_0) + \alpha_{ik}^{(2)}(T - T_0)^2 + \alpha_{ik}^{(3)}(T - T_0)^3$$

The expansion coefficients are all symmetric second order tensors and so in the general case have six independent components. Therefore up to third order an additional 18 quantities are required, making a grand total of $46 + 138 + 18 = 202$. Since in fact the temperature coefficients of the density are determined by the thermal expansion coefficients, the final count of the number of independent parameters required to specify the characteristics of a completely anisotropic material in the linear approximation reduces to 199.

Fortunately the completely anisotropic material is of little practical importance, and in commonly used materials the number of independent parameters is much reduced by the presence of crystal symmetries. In the particular case of quartz, with one trigonal and three digonal axes, it can be shown by the methods illustrated in Appendix 5 that the numbers of independent parameters for second, third and fourth rank tensor properties are 2, 2 and 6, respectively. Using the matrix notation also explained in Appendix 5, the dielectric, piezoelectric and elastic constants have the following forms.

MATERIAL CONSTANTS

Dielectric constants

$$\begin{pmatrix} \epsilon_{11} & 0 & 0 \\ 0 & \epsilon_{11} & 0 \\ 0 & 0 & \epsilon_{33} \end{pmatrix}$$

Piezoelectric constants

$$\begin{pmatrix} e_{11} & -e_{11} & 0 & e_{14} & 0 & 0 \\ 0 & 0 & 0 & 0 & -e_{14} & -e_{11} \\ 0 & 0 & 0 & 0 & 0 & 0 \end{pmatrix}$$

Elastic constants

$$\begin{pmatrix} c_{11} & c_{12} & c_{13} & c_{14} & 0 & 0 \\ c_{12} & c_{11} & c_{13} & -c_{14} & 0 & 0 \\ c_{13} & c_{13} & c_{33} & 0 & 0 & 0 \\ c_{14} & -c_{14} & 0 & c_{44} & 0 & 0 \\ 0 & 0 & 0 & 0 & c_{44} & c_{14} \\ 0 & 0 & 0 & 0 & c_{14} & c_{66} \end{pmatrix}$$

where $c_{66} = (c_{11} - c_{12})/2$.

Table 1.1 gives values for these independent parameters at a reference

Table 1.1 Material constants of right quartz

Property	Value at 25°C	First order, $10^{-6}/°C$	Temperature coefficients Second order, $10^{-9}/(°C)^2$	Third order, $10^{-12}/(°C)^3$
Thermal expansion				
α_{11}	—	+13.71	+ 6.5	−1.9
α_{33}	—	+ 7.48	+ 2.9	−1.5
Density (kg/m³)				
ρ	2648.6	−34.9	−15.9	+5.3
Elastic constants (10^9 N/m²)				
c_{11}	86.74	−48.5	−107	−70
c_{33}	107.20	−160	−275	−250
c_{12}	6.97	−3000	−3050	−1260
c_{13}	11.90	−550	−1150	−750
c_{44}	57.93	−177	−216	−216
c_{66}	39.89	+178	+118	+21
c_{14}	17.91	+101	−48	−590
Piezoelectric constants (C/m²)				
e_{11}	0.171	−160	—	—
e_{14}	0.0406	−1440	—	—
Dielectric constants (10^{-12} F/m)				
ϵ_{11}	39.97	—	—	—
ϵ_{33}	41.03	—	—	—

temperature of 25°C along with their temperature coefficients. There is reasonable agreement in the literature on the room temperature values but considerably less in respect of the temperature coefficients, with the uncertainty increasing as the order increases. In the case of the piezoelectric constants e_{11} and e_{14} even the first order coefficients are uncertain, and no information is available on the higher order coefficients of the dielectric constants ϵ_{11} and ϵ_{33}. The values presented in Table 1.1 are not intended to be definitive; for more detailed information on the material constants of quartz and references to the original papers recent review articles, such as that of Brice (1985), should be consulted.

1.4 MODES OF VIBRATION

Of the many different modes of vibration that may exist in a piezoelectric body, it is only those that are capable of being driven by an applied electric field that are of direct relevance in resonator applications. When the constitutive relations for a linear piezoelectric material are written in the form

$$S = sT + \tilde{d}E$$
$$D = dT + \epsilon E$$

where all quantities are understood to be expressed in the matrix form of Appendix 5, the first of these can be used to gain an understanding of the types of deformation that result from an applied field.

In the absence of mechanical stresses, the strain S is just $\tilde{d}E$ or in terms of the matrix components,

$$S_k = d_{jk} E_j$$

where $k = 1$ to 6 and j runs from 1 to 3, with the summation over j understood. The d matrix is subject to the same symmetry conditions and the matrix e, and so for the particular case of quartz has the form

$$\begin{pmatrix} d_{11} & -d_{11} & 0 & d_{14} & 0 & 0 \\ 0 & 0 & 0 & 0 & -d_{14} & -2d_{11} \\ 0 & 0 & 0 & 0 & 0 & 0 \end{pmatrix}$$

where the factor 2 in the component d_{26} arises from the corresponding factor of 2 in the definitions of the matrix components of strain S_4, S_5 and S_6 in Appendix 5. From the vanishing of the third row and third column of the d matrix it follows that an electric field component E_3 cannot produce any mechanical strain, and also that a longitudinal strain component S_3 cannot be produced by a field in any direction. (Note however that in the dynamic case where the stresses T are not zero, S_3 strains will be produced by elastic coupling due to the non-zero off-diagonal elastic constants.)

For field components E_1 and E_2, the possible non-zero strains are

$S_1 = d_{11}E_1$
$S_2 = -d_{11}E_1$
$S_4 = d_{14}E_1$
$S_5 = -d_{14}E_2$
$S_6 = -2d_{11}E_2$

Taking these in turn illustrates the basic modes of vibration commonly used in quartz resonators. In practical cases, these simple modes generally suffer from the disadvantages of high temperature coefficients of frequency, excessive unwanted coupling to other modes, or inconvenient dimensions in particular frequency ranges, and are consequently little used. However, they form the basis for the understanding of the more sophisticated resonators that have been developed for specific purposes and which are also briefly described in the following sections. Thickness mode resonators, which are currently the most important type in frequency control applications, are discussed in more detail in the following chapter. For further details of the other resonator types, reference should be made to the standard works by Heising (1946), Mason (1950) and Vigoreux and Booth (1950). A comprehensive review, especially with regard to low frequency units, is to be found in the US Government Report, *Handbook of Piezoelectric Crystals for Radio Equipment Designers* (Buchanan, 1956).

1.4.1 Thickness extensional modes in *X*-cut plates

The relation $S_1 = d_{11}E_1$ shows that an electric field along the *X* or electric axis produces a longitudinal strain along the same axis. Therefore if a plate is cut normal to the electric axis and provided with electrodes on its major faces, as in Fig. 1.2, a voltage applied to the electrodes will produce a strain along the thickness of the plate. An alternating voltage will therefore generate longitudinal waves in the plate with both the particle displacement and the direction of propagation along *X*. In a first approximation, mechanical resonance will occur when the frequency is such that the thickness of the plate contains an integral number n of half wavelengths of the longitudinal or extensional wave. Under such conditions, standing waves will be set up in the plate, with the plate surfaces being antinodes of vibration. With n odd the displacements of the two major surfaces are in anti-phase, whereas with n even the displacements are in phase. Consequently, as explained in more detail in the following chapter, the even order modes cannot be excited electrically because the applied potentials on the resonator electrodes are always in anti-phase. The integer n is the *overtone order*, and so in this case, and for thickness modes in general, only the odd overtones can be excited.

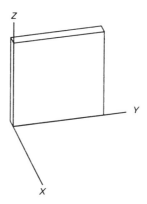

Fig. 1.2 X-cut plate.

The $n = 1$ mode is commonly termed the *fundamental* mode. Again to a first approximation, the velocity V of the extensional waves depends only on the material constants and is independent of the frequency f. The wavelength is (V/f), so the condition for resonance may be written

$$t = n(V/f)/2$$

or

$$f = nV/(2t)$$

where t is the thickness of the plate and n is an odd integer. Thus the resonance frequencies are inversely proportional to the thickness t. The product $k = ft$, known as the *frequency-thickness constant*, is $nV/2$ and so depends only on the material constants and the overtone order. In the present case of thickness extensional modes in X-cut plates, k has a value of approximately 2870 kHz mm and the temperature coefficient of frequency is about -20 ppm/°C. A plate of thickness 1 mm therefore has a frequency of approximately 2.87 MHz.

Besides the longitudinal strain S_1, an electric field along the electric axis also produces strains S_2 and S_4. S_2 is a longitudinal strain along the Y axis, which lies in the plane of an X-cut plate, and S_4 is a shear strain, again in the plane of the plate. Additionally, a strain S_3 is produced by elastic cross-coupling. All three strain types give rise to other families of resonances in X-cut plates, but with resonance frequencies essentially determined by the lateral dimensions and not the thickness. So long as the length and width of rectangular plates, and the diameter of circular plates, are kept much larger than the thickness, the fundamental thickness mode frequency will be much higher than the fundamental frequencies of these lateral modes, and will be essentially determined by the simple formula above. However, coupling between the thickness mode and higher order overtones of the lateral modes cannot be avoided, and this together with the large temperature coefficient effectively rules out the X-cut resonator from frequency-control applications. The X-cut nevertheless remains useful as an ultrasonic transducer.

1.4.2 Length extensional modes in X-cut bars

Even in the absence of the factors mentioned in the previous section which limit the use of X-cut thickness mode resonators, at frequencies below the megahertz region the plate dimensions would become too unwieldy for practical use. For such frequencies it becomes necessary to make use of those modes where the frequency is determined by dimensions other than the thickness. If instead of the plate in Fig. 1.2, the crystal blank is cut in the form of a bar with the length along the Y axis, as in Fig. 1.3, and width along Z, thickness along X as before, then a field E_1 applied via electrodes on the YZ faces can excite the bar into longitudinal vibration along Y via the relation $S_2 = -d_{11}E_1$. In an analogous way to the thickness mode case, an alternating field can be thought of as generating longitudinal waves propagating along Y with a velocity V essentially determined by the material constants and independent of the dimensions. Resonance occurs when an integral number of half wavelengths is included in the length l of the rod, so that if f is the frequency

$$f = nV/2l$$

For the X-cut bar, the frequency constant $k = fl$ is approximately 2730 kHz mm, so that for a length of 25 mm, the fundamental frequency is roughly 109 kHz. The frequency–temperature characteristic is parabolic in nature with a zero coefficient around room temperature but a total shift of around 100 ppm in a temperature range -20 to $+70°C$.

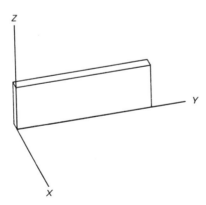

Fig. 1.3 X-cut bar.

Just as in the case of the X-cut plate, the additional strains S_1 and S_4 will also be excited. The S_1 strain, corresponding to the thickness mode, will have a fundamental mode frequency much higher than the length mode provided the thickness of the bar is sufficiently small, and outside the frequency range of interest. The S_4 strain, which is not only piezoelectrically driven but also coupled to the S_2 strain via the elastic compliance s_{24}, can, however, generate interfering modes of both face shear and flexural type. Interfering modes can also be generated in the width or Z direction through elastic cross coupling to the strain S_3, even though this strain is not piezoelectrically excited. Hence the width-to-length ratio for X-cut bars needs to be carefully chosen to avoid unwanted couplings in the desired frequency range.

1.4.3 Face shear modes in *X*-cut and *Y*-cut plates

The strain-field relations $S_4 = d_{14}E_1$ and $S_5 = -d_{14}E_2$ indicate that a shear in the plane of the plate is produced by a field along the thickness in both X-cut and Y-cut crystals. These shears are called *face shears* to distinguish them from *thickness shears* in which the deformation occurs in a plane containing the thickness of the crystal, as illustrated in Fig. 1.4. A static face shear would deform an initially square plate into a rhombus; however, the Figure shows in schematic form the deformation expected in the dynamic case when the driving field is an alternating one. In contrast to the previously considered cases of extensional modes along either the length or thickness of a plate, and to the case of thickness shear modes to be considered next, there is no comparable simple explanation of face shear modes in terms of one-dimensional standing wave systems. This is due to their essentially two-dimensional nature. For rectangular plates of length l and width w, the face shear resonance frequencies depend upon both l and w, and the overtone frequencies are not harmonically related. Because of the strong coupling that exists to other modes and their unfavourable temperature coefficients, face

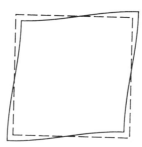

Fig. 1.4 Face shear deformation of a square plate.

shear modes in X- and Y-cut plates are not used in practice; however, in oblique cuts, ie, plates cut at an angle to the crystallographic axes, the face shear mode is very important.

1.4.4 Thickness shear modes in Y-cut plates

The strain-field relation $S_6 = -2d_{11}E_2$ shows that a shear strain in the XY plane is developed by a field along Y. Since the strain involves the thickness direction, it is termed a *thickness shear* as opposed to the face shears considered in the previous section. As shown in the following chapter, the principal displacement in the thickness shear deformation of a Y-cut plate is in the X direction and therefore perpendicular to the thickness (Fig. 1.5). An alternating field can thus be thought of as generating transverse waves propagating along the thickness with a velocity V dependent in a first approximation only on the material constants, and just as in the case of thickness extensional modes in X-cut plates, resonance will occur when the thickness contains an integral number of half wavelengths. The resonance frequencies will again be given by the expression

$$f = nV/2t$$

and a frequency-thickness constant $k = ft = nV/2$ can be defined. Once again, only odd order overtones can be electrically excited.

For the Y-cut resonator, k is approximately 1950 kHz mm and the temperature coefficient of frequency is about +75 ppm/°C. Not only is the temperature coefficient rather large, but the actual frequency–temperature

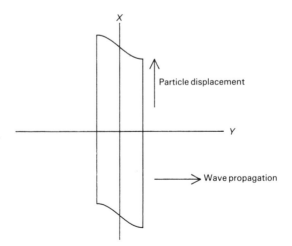

Fig. 1.5 Thickness shear deformation in Y-cut plate.

characteristic is subject to large discontinuities as a result of the very strong coupling to overtones of the face shear and other interfering modes. Consequently the *Y*-cut is no longer used in frequency control applications, although it still finds application as an ultrasonic transducer for the generation of shear waves.

1.4.5 Flexural modes

The modes of vibration discussed so far all have direct links with the matrix strain-field relationship $S = \tilde{d}E$ insofar as the principal strain involved is one of those excited by a field component along a crystallographic axis. Flexural modes need a rather more complex electrode arrangement for their piezoelectric excitation, but in themselves represent probably the most obvious and widespread vibration type. Everyday examples are provided by tuning forks, xylophones, and all types of wind instruments dependent on vibrating reeds. The major advantage of flexural modes in the context of frequency control is that in the low frequency range, say below about 50 kHz, other types of resonator become unmanageably large, whereas the flexural type can be kept compact. This is illustrated admirably by the modern application in electronic watches of miniature crystals at 32.768 kHz utilizing flexural modes in a tuning fork structure (cf. the paper by Walls in Gerber and Ballato, 1985, **2**, p. 276, for a brief review and references).

Length extensional vibrations in an *X*-cut bar with its length along the *Y* axis have already been considered in Section 1.4.2. Such a bar can be excited in a flexural mode by reversing the field direction over half the width of the bar, so that the upper half contracts while the lower half expands and vice versa (Fig. 1.6). The fundamental mode of such a resonator has two nodal points, as shown in Fig. 1.7, with the middle and both ends of the bar being anti-nodes. This is termed a *free-free* mode in contrast to a *clamped-free* mode where one end of the bar is clamped and thus constrained to be a node, as in a tuning fork. A further type of flexural resonator, of importance in the lowest frequency ranges, is the *bimorph* or *duplex* resonator. This consists of

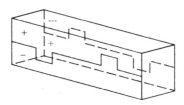

Fig. 1.6 Excitation of flexural modes in a bar.

Fig. 1.7 Flexural deformation of a bar.

two length extensional bars of opposite hand cemented together such that one bar expands while the other contracts. The frequency range covered is from a few hundred Hz up to about 10 kHz.

For all types of flexural mode resonator, the resonance frequencies are in general complicated functions of the dimensions. In particular, it is not possible except in special limiting cases to define a useful frequency constant, and the overtone mode frequencies are not harmonically related to the fundamental.

1.4.6 Coupled mode resonators: the GT-cut

The crystal resonators so far described have all been in the form of bars or plates cut along or normal to the crystallographic axes. They suffer either from large temperature coefficients of frequency, or severe unwanted coupling to other modes of vibration, or both. As such, although offering significant advances in frequency stability as compared to previous techniques, the performance of early crystal units utilizing these cuts still fell short of the requirements of radio communications systems. Noting that the temperature coefficients of the simple cuts had both positive and negative values, early workers in the field were led to investigate two avenues of approach. In the first, the objective was to produce a resonator that by virtue of coupling between two modes of motion with different temperature coefficients, would have a zero resultant temperature coefficient. In the second, the objective was to find orientations such that plates cut accordingly would have a zero temperature coefficient for a simple uncoupled mode of vibration.

The culmination of the work in the first direction was the invention by Mason (1940) of the GT-cut resonator. This rests on the facts that (a) plate orientations can be found such that the face shear mode of vibration has a positive temperature coefficient; that (b) a face shear motion can be resolved into two extensional motions at right angles; and that (c) it can be shown that all pure extensional modes have negative or zero temperature coefficients. It follows that the positive temperature coefficient of the face shear mode must be due to the coupling between the two component extensional modes, and Mason recognized that by reducing this coupling the coefficient of the original mode must eventually pass through zero.

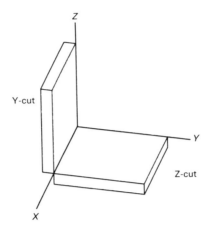

Fig. 1.8 Y-cut and Z-cut plates.

The procedure by which the coupling is controlled can be explained as follows. Consider first a square Y-cut plate, that is a plate cut normal to the Y axis with its length and width along the X and Z axes. As described in Section 1.4.3, such a plate may be driven in a face shear mode by a field along Y through the d_{14} piezoelectric constant. The temperature coefficient of the face shear mode is negative, but it turns out that as the plate is rotated about the X axis towards the limiting case of a Z-cut plate (Fig. 1.8), the temperature coefficient increases through zero to a positive value. (At the same time the degree of piezoelectric excitation decreases to zero, as the component of the applied field along the crystallographic Y axis decreases and that along the non-piezoelectric Z axis increases.) Suppose that an orientation is selected such that the face shear mode has a positive temperature coefficient, but that the plate is cut with its length at 45° to the X axis. Then the face shear character of the piezoelectrically driven motion will disappear and be replaced by a pair of coupled extensional vibrations along the length and width of the rotated plate (Fig. 1.9). These vibrations will be close in

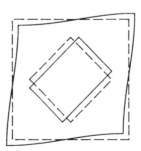

Fig. 1.9 Face shear and coupled extensional modes.

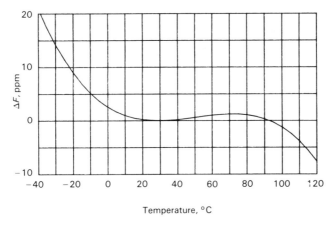

Fig. 1.10 GT-cut frequency–temperature characteristic.

frequency to that of the face shear mode and at least one will have a positive temperature coefficient. If now either the length or the width of the plate is altered, the coupling between the extensional modes will be reduced as the separation between the uncoupled mode frequencies is increased. As the coupling is further and further reduced by altering the width-to-length ratio of the plate, then the temperature coefficients of the two modes must both approach those of the pure extensional modes, which are negative. Hence at one particular value of the width-to-length ratio, the temperature coefficient of one of the coupled modes must vanish. As it turns out, by judicious choice of the angle of rotation about the X axis and also of the dimensional ratio, both first and second order temperature coefficients of frequency can be made to vanish at room temperature, resulting in a frequency–temperature characteristic extremely flat over a range of 100°C or more (Fig. 1.10).

The GT-cut resonator was originally conceived by Mason as one of a family of similar cuts varying in rotation angle about X and width-to-length ratio. The optimum values given by Mason for the temperature range $-25°$C to $+75°$C are a rotation angle of $-51°$ 7.5' and a width-to-length ratio of 0.859. In this case the zero temperature coefficient mode is the width extensional one, with a frequency constant of 3370 kHz mm, so that the frequency can be adjusted by modifying the width while small corrections to the temperature coefficient can be independently made by adjusting the length. The subsidiary length extensional mode has a frequency some 15% lower than the main width mode.

Despite its excellent frequency stability, early application of the GT-cut was mainly limited to frequency standards in the frequency range 100 to 500 kHz, primarily because of manufacturing difficulty. To achieve maximum performance requires extreme care and precision in the orientation and dimensioning of the crystal blank, accompanied by equal care in mounting. Compared to the relative ease of manufacture of AT-cut plates, the high manufacturing costs of the GT could only be justified in rare cases. In the last few years, however, this situation has been changed by the application of photolithographic manufacturing techniques and computer-aided design to the production of miniature GT resonators in the frequency range 1 to 3 MHz. An interesting feature of these units is the fabrication of both resonator and support structure out of a single quartz blank (Kawashima et al., 1980).

1.4.7 Rotated crystal cuts

The second avenue of approach to the objective of improved frequency stability was, as mentioned in the previous section, to search for orientations such that crystals cut accordingly had favourable characteristics in respect of either freedom from unwanted coupled modes, or frequency–temperature coefficient or both. The pioneering research in this area was mainly done by a team of workers at the Bell Telephone Laboratories, and is fully described in the books by Heising (1946) and Mason (1950).

The basic theory and results for thickness mode resonators in arbitrarily oriented plates are considered in Chapter 2, and it is shown there that the first-order temperature coefficient of frequency of such resonators is given by

$$T_f = \tfrac{1}{2} T_c - \tfrac{1}{2} T_\rho - T_t$$

where T_c, T_ρ and T_t are, respectively, the temperature coefficients of the effective elastic constant, the density and the thickness. Both T_ρ and T_t are determined by the thermal expansion coefficients, and it can easily be shown (Chapter 2) that their combined contribution to T_f is always small and positive, ranging from a minimum of about 4 ppm/°C to a maximum of 10 ppm/°C. As the first-order coefficients of the elastic constants listed in Table 1.1 are typically an order of magnitude greater than this, it is clear that T_f is primarily determined by T_c.

The *rotated Y-cut* family of plates can be thought of as obtained from a Y-cut plate by rotation about the crystallographic X axis, which lies in the plane of the plate. For all plates in this family, it can be shown (Chapter 2) that the only one of the three possible thickness modes that can be piezoelectrically excited is a thickness shear mode with the particle displacement along the X axis, with the effective elastic constant c_{66}'. The prime indicates that the indices refer to a plate coordinate system with X axis coincident with the crystal X axis, Y axis along the thickness of the plate and the Z axis chosen to form a right-handed system. As shown in Appendix 5, c_{66}' is then given in terms of the rotation angle θ by

$$c_{66}' = s^2 c_{44} + 2sc\, c_{14} + c^2 c_{66}$$

where $s = \sin(\theta)$ and $c = \cos(\theta)$. Hence as θ varies between 0 and 90° c_{66}' varies from c_{44} to c_{66}. But Table 1.1 shows that the temperature coefficients of c_{44} and c_{66} are roughly equal in magnitude but opposite in sign, so that the temperature coefficient of c_{66}' must vary continuously from large negative values to large positive values. Therefore there must be at least one angle θ for which T_c just compensates for the contributions of T_ρ and T_t to give a zero T_f.

In fact there are two such angles when positive and negative rotations are considered, corresponding to the AT- and BT-cuts first reported by Lack, Willard and Fair (Lack et al., 1934). The corresponding angles are approximately $-35.25°$ and $+49.20°$ for AT and BT, respectively, the precise angles being varied in manufacture to suit the requirements of particular specifications. The actual frequency-temperature characteristics of the AT and BT are, respectively, found to be cubic and parabolic in nature, with the AT-cut having the distinct advantage that both first *and* second order temperature coefficients can be made to vanish around room temperature, thus giving a flatter frequency-temperature curve over a wider range of temperature than the BT.

As remarked in Section 1.4.4, one of the problems with the simple Y-cut thickness shear resonator is the strong coupling with overtones of the face shear mode involving an S_5 strain. Not only is the face shear mode directly driven through the d_{14} piezoelectric constant, it is also coupled to the thickness shear motion through the c_{56} elastic constant. This coupling is also present in the rotated Y-cuts. Again referring to Appendix 5, in the plate coordinate system the constant c_{56}' is determined in terms of the rotation angle by

$$c_{56}' = c_{14}(c^2 - s^2) - (c_{66} - c_{44}) sc$$

This is easily shown to vanish when

$$\tan(2\theta) = 2c_{14}/(c_{66} - c_{44})$$

that is, for rotation angles of $-31.6°$ and $+58.4°$. Hence at these angles, known respectively as the AC- and BC-cuts and also introduced by Lack, Willard and Fair, the mechanical coupling to the face-shear mode vanishes. As the AC and BC angles are quite close to the AT and BT, it also turns out that the latter cuts are also relatively free from this coupling. However they do remain prone to coupling to flexural modes.

The AC, BC, AT and BT are the most important thickness mode resonators in the rotated Y-cut family, but are not suitable for low frequency applications, say below 1 MHz. In the frequency region down to about 200 kHz, the face shear resonators designated the CT- and DT-cuts and introduced by Hight and Willard (1937) are commonly used. These are closely related to the AT- and BT-cuts as shown in Fig. 1.11 for the AT and DT case. It is clear from the figure that the face shear motion in the DT plate involves the same strain component S_6' as does the thickness shear motion in

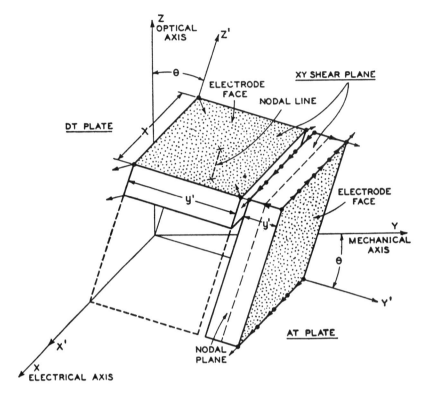

Fig. 1.11 Relation of AT and DT crystal cuts. (Reproduced, with permission, from R.A. Heising, *Quartz Crystals for Electronic Circuits*, Electronic Industries Association, 1978.)

the AT plate, and should therefore be governed by the same elastic constant, c_{66}'. Since this has a near zero temperature coefficient, it is also to be expected that the face shear mode will have a near zero frequency–temperature coefficient. Willard and Hight found in practice that this was achieved with a difference in angle of 87° rather than the 90° of the figure, corresponding to a rotation angle for the DT-cut of +52°. The CT is similarly related to the BT, and has a rotation angle of −38°.

The manufacture of crystal blanks in the rotated Y-cut family is relatively straightforward. In the initial cutting or sawing operations and in the subsequent lapping and polishing processes, precise control is necessary of the angle by which the plate is rotated about the X axis. Tolerances on this angle of the order of a minute of arc are common, but on the other hand the tolerances on the incidental rotations that may also occur about the Y and Z axes are much wider, with errors of the order of a degree being quite tolerable.

Such a situation no longer exists when the performance requirements imposed on the crystal are such as to demand more complicated cuts. The

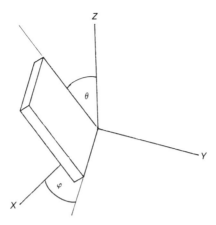

Fig. 1.12 Doubly rotated crystal plate.

AT-cut, for example, although the most widely used thickness mode resonator, has certain disadvantages in relation to its performance when mechanically and thermally stressed. These can be overcome by introducing the extra degree of freedom allowed by the use of *doubly rotated* cuts. As illustrated in Fig. 1.12, a doubly rotated cut can be thought of as being obtained from a Y-cut plate by an initial rotation about the Z axis through an angle ϕ, followed by a second rotation about the X' axis through an angle θ, so that the angles ϕ, θ specify the orientation of the blank. In the special case of the rotated Y-cuts, the first rotation ϕ is zero.

The benefit of allowing ϕ to vary over its permissible range is that a whole family of crystal cuts having a zero frequency–temperature coefficient at or around room temperature is obtained. In fact two distinct families of zero coefficient crystals are generated, one containing the AT as a special case, the other containing the BT. Particularly extensive investigations of the properties of the family containing the AT-cut have been carried out, culminating in the introduction of the *stress-compensated*, or SC-cut, with $\phi \sim 22°$ and $\theta \sim -34.3°$ (EerNisse, 1975).

The specific advantages of the SC are that for a plate so oriented, the resonance frequencies of the plate are largely independent of mechanical stresses in the plane of the plate. Such stresses can arise from external forces, from the mounting of the crystal, from stresses in the electrode films, and also from thermal gradients set up in the crystal when subject to varying ambient temperatures. Frequency shifts due to these causes are relatively large in AT-cut resonators and limit the frequency stability available. Consequently, the SC-cut is currently preferred for those applications where the highest stability is required.

As already intimated, the advantages of the doubly rotated cuts are accompanied by significant disadvantages. Firstly, the manufacturing difficulties

Table 1.2 Rotated X-cuts

Cut	Mode of vibration	Frequency range, kHz
X	Thickness extensional	350 to 20 000
X	Length extensional	40 to 350
$+5°X$	Length extensional	50 to 500
$+5°X$	Length-width flexure	10 to 100
$+5°X$	Duplex	0.4 to 10
MT	Length extensional	50 to 500
NT	Length-width flexure	4 to 100
$-18°X$	Length extensional	60 to 300

are considerably complicated by the need to maintain close tolerances on both rotation angles. Secondly, whereas in the rotated Y-cut case only a single thickness mode is excited, in the general doubly rotated cases all three modes can be excited. For the SC, the second thickness-shear mode is nearly equal in strength to the desired mode, and only some 10% above it in frequency. This makes it necessary to either use additional circuitry to suppress the unwanted mode, or else to use special manufacturing techniques to reduce its relative strength. The advantages and disadvantages of doubly rotated cuts are considered in detail by Ballato (1977).

To complete the discussion on rotated crystal cuts, brief mention must be made of the crystals in the *rotated* X-*cut* family that are used for low frequency applications. Table 1.2 lists the commonly used cuts, giving their usual designations, the mode of vibration and the frequency range in which they operate. As for the thickness mode resonators already discussed, the motivation for introducing these rotated cuts is to improve the frequency-temperature characteristics or to reduce interference from unwanted coupled modes, or both. For a full discussion, reference should be made to Heising (1946) or Buchanan (1956).

2 Thickness mode resonators

2.1 WAVE PROPAGATION IN PIEZOELECTRIC MATERIALS

The subject of the propagation of elastic waves in solid materials has a long history, although the extension of the analysis to give a complete treatment in the piezoelectric case has only been made relatively recently. The book *Linear Piezoelectric Plate Vibrations* by Tiersten (1969), and the review article *Doubly Rotated Thickness Mode Plate Vibrators* by Ballato (1977) should be consulted for references to the earlier work.

The field equations and constitutive relations for a linear piezoelectric material are (Appendix 4)

$$t_{kl,l} = \rho \ddot{u}_k \tag{2.1}$$

$$D_{k,k} = 0 \tag{2.2}$$

$$E_k = -\phi_{,k} \tag{2.3}$$

$$t_{kl} = c_{klmn} S_{mn} - e_{mkl} E_m \tag{2.4}$$

$$D_k = e_{klm} S_{lm} + \epsilon_{kl} E_l \tag{2.5}$$

where ρ is the mass density, u_k the particle displacement, E_k, D_k and ϕ are the electric field, displacement and potential respectively, t_{kl} and S_{kl} are the elastic stress and strain, and c_{klmn}, e_{klm} and ϵ_{kl} are the elastic, piezoelectric and dielectric constants. For any function f, differentiation with respect to the position coordinate x_k is denoted by $f_{,k}$, whereas the partial derivative with respect to the time is written $\dot{f}$. The strain S_{kl} is defined in terms of the displacements u_k by

$$S_{kl} = (u_{k,l} + u_{l,k})/2 \tag{2.6}$$

Using Eqns (2.3) to (2.6), the field equations (2.1) and (2.2) can be expressed in terms of the displacements u_k and the potential ϕ by

$$c_{klmn} u_{m,nl} + e_{mkl} \phi_{,ml} = \rho \ddot{u}_k \tag{2.7}$$

$$e_{klm} u_{l,mk} - \epsilon_{kl} \phi_{,lk} = 0 \tag{2.8}$$

For solutions to Eqns (2.7) and (2.8) in the form of plane waves propagating along an arbitrary direction specified by a unit vector n_k, the field variables reduce to functions of the distance $x = n_k x_k$ measured along n_k. Then for any field f, the spatial derivatives $f_{,k}$ are given by $n_k f_{,x}$, where $f_{,x}$ denotes the derivative with respect to x. Consequently, Eqns (2.7) and (2.8) become

$$L_{km} u_{m,xx} + e_k \phi_{,xx} = \rho \ddot{u}_k \tag{2.9}$$

$$e_m u_{m,xx} - \epsilon \phi_{,xx} = 0 \tag{2.10}$$

with the definitions

$$L_{km} = c_{klmn} n_l n_n \tag{2.11}$$

$$e_l = e_{klm} n_k n_m \tag{2.12}$$

$$\epsilon = \epsilon_{kl} n_k n_l \tag{2.13}$$

Equation (2.10) can immediately be integrated to give the potential in terms of the displacements

$$\phi = e_k u_k / \epsilon + Cx + D \tag{2.14}$$

where C and D are arbitrary functions of the time t but independent of x. Also, Eqn (2.10) can be used to eliminate ϕ from Eqn (2.9), leading to

$$\bar{L}_{km} u_{m,xx} = \rho \ddot{u}_k \tag{2.15}$$

$$\bar{L}_{km} = L_{km} + e_k e_m / \epsilon \tag{2.16}$$

The surface tractions across a surface normal to n_k are given by $t_k = t_{kl} n_l$, and the normal component of the electric displacement is $D_k n_k$. From the constitutive relations (2.4) and (2.5)

$$t_k = L_{km} u_{m,x} + e_k \phi_{,x}$$

$$D_k n_k = e_l u_{l,x} - \epsilon \phi_{,x}$$

Substituting for ϕ from Eqn (2.14) gives

$$t_k = \bar{L}_{km} u_{m,x} + e_k C \tag{2.17}$$

$$D_k n_k = -\epsilon C \tag{2.18}$$

Because of the symmetry of the elastic constants expressed by the relation $c_{klmn} = c_{mnkl}$, the L_{km}, and therefore also the $\bar{L}_{km}$, are the components of symmetric second rank tensors. As shown in Appendix 4, the form $L_{km} x_k x_m$ is positive definite, and since

$$\bar{L}_{km} x_k x_m = L_{km} x_k x_m + (e_k x_k)^2 / \epsilon$$

it follows that $\bar{L}_{km} x_k x_m$ is also positive definite. Hence (Section A2.6) a coordinate system can be found in which $\bar{L}_{km} = \delta_{km} c_m$ (no sum over m) and in which all the eigenvalues c_m are positive. In this special coordinate system, the wave equation, Eqn (2.15), separates into three uncoupled equations for the three displacement components

$$c_k u_{k,xx} = \rho \ddot{u}_k \quad \text{(no sum on } k\text{)} \tag{2.19}$$

For each k, $k = 1, 2$ and 3, Eqn (2.19) is an elementary one-dimensional wave equation, representing waves propagating along x with a phase velocity $V_k = (c_k/\rho)^{1/2}$ and c_k representing an effective elastic constant. Clearly, by the choice of coordinate system, the particle displacement for the wave with velocity V_k is along the x_k axis, so that the particle displacements for the three waves are mutually orthogonal.

This is strictly true only when the phase velocities V_k are all unequal. In the degenerate case when two of the velocities have equal magnitudes, then clearly any linear combination of the corresponding displacements will also propagate with the same velocity; nevertheless, it is always possible to resolve any such waves into two components with mutually perpendicular displacements.

Physically, the preceding analysis means that for an arbitrary direction in an elastic or piezoelectric material, there will in general be three possible types of plane wave, distinguished by different phase velocities and mutually perpendicular particle displacements. In the special case of an isotropic material, and also in certain directions of high symmetry in crystalline materials, the coordinate system that diagonalizes $\bar{L}_{km}$ is such that the propagation direction lies along one of the axes. The three wave types can then be classified as longitudinal and transverse, the longitudinal wave being characterized by the particle displacement lying along the direction of propagation, whereas the two transverse waves have the particle displacement at right angles to the propagation direction. Generally, however, this will not be the case, and the longitudinal/transverse classification will at best serve as an approximate indication of the wave character.

2.2 BOUNDARY CONDITIONS FOR THICKNESS MODES

Thickness mode resonators are typically cut in the form of thin, flat plates whose lateral dimensions are much greater than the thickness. In a first approximation, the analysis of such resonators is based on the physical idea of a system of standing waves set up in the resonator by plane waves propagating along the thickness and being reflected at the major surfaces of the plate. The lateral dimensions are assumed to be effectively infinite. Of course, this is an oversimplification of the real problem, and the effect of finite lateral dimensions (considered in Chapter 3) cannot be neglected in practice. Nevertheless, the pure thickness mode approximation provides a great deal of insight and understanding, particularly in the context of the properties of oblique crystal cuts.

As shown in the previous section, for each propagation direction three types of plane wave exist. In a resonator therefore it is to be anticipated that there will be three distinct standing wave systems, which may or may not be coupled by the boundary conditions at the surfaces of the plate. Consider a resonator with major surfaces at $x = \pm h$ coated with massless, conducting electrodes that are connected to a voltage source of emf E. Then appropriate electrical boundary conditions are that the potential $\phi = \pm E/2$ at $x = \pm h$. Since the electrodes are assumed massless, the mechanical boundary conditions are simply that the surface tractions $t_k = t_{kl} n_l$ vanish at $x = \pm h$. Using Eqns (2.14) and (2.17) these conditions are

$$\phi(h) = e_k u_k(h)/\epsilon + Ch + D = E/2 \qquad (2.20)$$

$$\phi(-h) = e_k u_k(-h)/\epsilon - Ch + D = -E/2 \qquad (2.21)$$

$$t_k(h) = \bar{L}_{km} u_{m,x}(h) + e_k C = 0 \qquad (2.22)$$

$$t_k(-h) = \bar{L}_{km} u_{m,x}(-h) + e_k C = 0 \qquad (2.23)$$

Adding and subtracting Eqns (2.20) and (2.21) gives two equivalent equations

$$e_k[u_k(h) + u_k(-h)]/\epsilon + 2D = 0 \qquad (2.24)$$

$$e_k[u_k(h) - u_k(-h)]/\epsilon + 2Ch = E \qquad (2.25)$$

Similarly, adding and subtracting Eqns (2.22) and (2.23),

$$\bar{L}_{km}[u_{m,x}(h) + u_{m,x}(-h)] + 2e_k C = 0 \qquad (2.26)$$

$$\bar{L}_{km}[u_{m,x}(h) - u_{m,x}(-h)] = 0 \qquad (2.27)$$

Writing u_k as the sum of its symmetric and antisymmetric parts u_k^S and u_k^A and noting that the derivative of the symmetric part $u_{k,x}^S$ is antisymmetric in x and vice versa, Eqns (2.24) to (2.27) may be rewritten

$$e_k u_k^S(h)/\epsilon + D = 0 \qquad (2.28)$$

$$e_k u_k^A(h)/\epsilon + Ch = E/2 \qquad (2.29)$$

$$\bar{L}_{km} u_{m,x}^A(h) + e_k C = 0 \qquad (2.30)$$

$$\bar{L}_{km} u_{m,x}^S(h) = 0 \qquad (2.31)$$

with the definitions

$$u_k^S(x) = [u_k(x) + u_k(-x)]/2$$
$$u_k^A(x) = [u_k(x) - u_k(-x)]/2$$

Equations (2.28) to (2.31) reveal that the symmetric and antisymmetric parts of u_k are not coupled by the boundary conditions on the surface of the plate, and moreover that the symmetric part is not driven by the applied emf E, This is the fundamental reason why the symmetric even order overtones of thickness mode resonators cannot be electrically excited in normal circumstances (Section 1.4.1). In determining the resonance frequencies,

2.3 RESONANCE FREQUENCIES AND ELECTROMECHANICAL COUPLING

The field equations and boundary conditions for antisymmetric functions u_k, when stated in the coordinate system for which $\bar{L}_{km}$ is diagonal, are

$$c_k u_{k,xx} = \rho \ddot{u}_k \quad \text{(no sum on } k\text{)} \tag{2.19}$$

$$e_k u_k(h)/\epsilon + Ch = E/2 \tag{2.29a}$$

$$\bar{L}_{km} u_{m,x}(h) + e_k C = 0 \tag{2.30a}$$

where by assumption $u_k = u_k^A$. Appropriate solutions to Eqn (2.19) are

$$u_k(x,t) = A_k \sin(\beta_k x) \exp(j\omega t) \quad \text{(no sum on } k\text{)} \tag{2.32}$$

where the wavenumbers β_k are given by

$$c_k \beta_k^2 = \rho \omega^2 \quad \text{(no sum on } k\text{)} \tag{2.33}$$

Writing C and E in the form $C_0 \exp(j\omega t)$ and $E_0 \exp(j\omega t)$, and cancelling the common factor $\exp(j\omega t)$, the boundary conditions become

$$\sum_k \left\{ e_k A_k \sin(X_k)/\epsilon \right\} + C_0 h = E_0/2 \tag{2.34}$$

$$\sum_k \left\{ \bar{L}_{km} A_m B_m \cos(X_m) \right\}' + e_k C_0 = 0 \tag{2.35}$$

where X_k has been written for $(\beta_k h)$. Since $\bar{L}_{km}$ is diagonal, Eqn (2.35) may be simplified to

$$c_k X_k A_k \cos(X_k) + e_k C_0 h = 0 \tag{2.36}$$

with there being no sum on k. Solving for the A_k gives

$$A_k = -e_k C_0 h / [c_k X_k \cos(X_k)] \tag{2.37}$$

and substituting in Eqn (2.34) gives a solution for C_0 in terms of the driving voltage E_0:

$$C_0 = (E_0/2h) / \left\{ 1 - \sum_m [k_m^2 \tan(X_m)/X_m] \right\} \tag{2.38}$$

where the *electromechanical coupling factor* for the mth mode, k_m, is defined by

$$k_m^2 = e_m^2 / \epsilon c_m \quad \text{(no } m \text{ sum)} \tag{2.39}$$

The current density in the resonator is given by $J = -j\omega D_k n_k$ and therefore from Eqn (2.18)

$$J = j\omega\epsilon C_0 = j\omega(\epsilon/2h)E_0/\left\{1 - \sum_m [k_m^2 \tan(X_m)/X_m]\right\} \quad (2.40)$$

Since $c_0 = (\epsilon/2h)$ is just the static capacitance of the plate per unit area, and E_0 is the applied voltage, the admittance y per unit area of the plate is given by

$$y = j\omega c_0/\left\{1 - \sum_m [k_m^2 \tan(X_m)/X_m]\right\} \quad (2.41)$$

The zeros and poles of the admittance correspond to the anti-resonance and resonance frequencies, respectively. The zeros coincide with the poles of the tangent functions in Eqn (2.40), and hence occur when $\cos(X_m) = 0$, or $X_m = M\pi/2$ where M is an odd integer. The corresponding frequencies are, from Eqn (2.33)

$$\begin{aligned}\omega &= (c_m/\rho)^{1/2}M\pi/2h\\ f &= (c_m/\rho)^{1/2}M/4h = MV_m/4h\end{aligned} \quad (2.42)$$

where V_m is the phase velocity of the mth mode. There are thus three harmonically related sequences of anti-resonance frequencies; if necessary, the notation $f_{mA}^{(M)}$ can be used to discriminate between them (Ballato, 1977).

The series resonance frequencies are determined by the zeros of the denominator of y, that is by the roots of the equation

$$\sum_m k_m^2 \tan(X_m)/X_m = 1 \quad (2.43)$$

Unlike the case of the anti-resonance frequencies, Eqn (2.43) does not generally allow the identification of separate sequences of roots each associated with a particular displacement mode. In the general case, for a given root of Eqn (2.43), all three modes will be excited. Obvious exceptions occur when one or two of the k_m vanish, since then the corresponding modes cannot be piezoelectrically excited. A very important case is that of the rotated Y-cut family, where only one k_m is non-zero, and only a single thickness shear mode excited. Even when two or three modes are excited, it may happen that conditions are such that one term in the sum in Eqn (2.43) dominates, so that as an approximation the other terms can be neglected. A detailed discussion will be found in Ballato (1977); the single mode case is briefly considered here in order to arrive at the important relationship between the coupling factor and the separation of the resonance and anti-resonance frequencies.

In the single mode case, Eqn (2.43) can be rewritten, dropping the suffices, as

$$\tan(X) = X/k^2 \quad (2.44)$$

Since for quartz, for all modes and all propagation directions, $k^2 \ll 1$, it

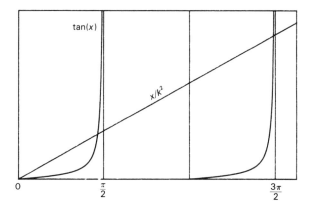

Fig. 2.1 Graphical solution of $\tan(X) = X/k^2$.

follows that the straight line X/k^2 has a much steeper slope than the tangent curve near the origin, as illustrated in Fig. 2.1. Therefore Eqn (2.44) will always have a root in the interval $[0, \pi/2]$, and an infinite sequence of roots thereafter, one in each of the intervals $[N\pi, (N + 1/2)\pi]$, for $N = 1, 2, 3. \ldots$ It is clear from the figure that as N increases, the roots tend to the poles of the tangent function, and that for fixed N, the separation between the root and the corresponding pole decreases as k decreases. Since as already pointed out, the poles of the tangent function correspond to the anti-resonance frequencies, the separation between resonance and anti-resonance frequencies diminishes with increasing N and decreasing k. Let $M = 2N + 1$ and let $X_R^{(M)}$ denote the root of Eqn (2.44) that lies in the interval $[N\pi, M\pi/2]$. Then if $X_A^{(M)} = M\pi/2$, and the fractional frequency difference $\delta_R^{(M)}$ is defined by

$$\delta_R^{(M)} = (X_A^{(M)} - X_R^{(M)})/X_A^{(M)} \tag{2.45}$$

$\delta_R^{(M)}$ is always $\ll 1$.

If for a general X, a corresponding δ is defined by

$$\delta = (X_A^{(M)} - X)/X_A^{(M)} \tag{2.46}$$

so that $X = X_A^{(M)}(1 - \delta)$, then $\tan(X) = \cot(\delta X_A^{(M)})$ and Eqn (2.44) can be rewritten as an equation in δ

$$k^2 = X_A^{(M)}(1 - \delta)\tan(\delta X_A^{(M)}) \tag{2.47}$$

For small δ the tangent can be replaced by its argument, and then to first order in δ, Eqn (2.47) has the solution

$$\delta_R^{(M)} = \delta = 4k^2/M^2\pi^2 \tag{2.48}$$

(Note: the fractional frequency difference defined above differs by a factor M from the *frequency displacement* defined by Ballato (Ballato, 1977))

2.4 EQUIVALENT CIRCUITS

Since the admittance y per unit area of the plate as defined by Eqn (2.41) has an infinite number of poles and zeros, no exact finite lumped parameter equivalent circuit exists. Equivalent circuits in the form of lengths of transmission lines have been proposed, but in the context of resonator theory and applications are unnecessarily cumbersome. For most practical purposes, it is entirely adequate to concentrate attention on the neighbourhood of a particular resonance frequency and employ a simple lumped element circuit valid only in that neighbourhood. Considering then the Mth order resonance of a plate in which only a single mode is excited (only one k_m non-zero), the admittance Y of an electrode patch of area A can be written in terms of the fractional frequency difference δ as

$$Y = j\omega C_0/\{1 - k^2 \tan(X)/X\}$$
$$= j\omega C_0 X_A^{(M)} (1 - \delta)\tan(\delta X_A^{(M)})/\{X_A^{(M)} (1 - \delta)\tan(\delta X_A^{(M)}) - k^2\}$$

where $C_0 = Ac_0 = A\epsilon/2h$. For small δ, this reduces to

$$Y = j\omega C_0 \delta/(\delta - \delta_R^{(M)}) \tag{2.49}$$

As anticipated, this expression has a single zero and a single pole, just as does the admittance, of the simple circuit of Fig. 2.2. The analysis of Section 6.2.1, shows that the fractional frequency difference between the pole and zero of Fig. 2.2 is determined by the capacitance ratio C_0/C_1. Therefore the equivalence is assured provided that

$$C_1/2C_0 = 4k^2/M^2\pi^2 \tag{2.50}$$

Since C_0 is known in terms of the electrode area, plate thickness and the effective dielectric constant ϵ, Eqn (2.50) determines C_1 in terms of the plate parameters. The inductance L_1 in Fig. 2.2 is of course determined by C_1 and the resonance frequency ω_r through $L_1 = 1/\omega_r^2 C_1$. Since the separation between resonance and anti-resonance is small, the simple expression Eqn (2.42) for the anti-resonance frequencies can be used in estimating L_1.

Summarizing, the parameters in the equivalent circuit are given by

$$\left. \begin{array}{l} C_0 = \epsilon A/2h \\ C_1 = 8k^2 C_0/M^2\pi^2 = 4e^2 A/(chM^2\pi^2) \\ L_1 = \rho h^3/(e^2 A) \end{array} \right\} \tag{2.51}$$

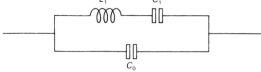

Fig. 2.2 Lossless equivalent circuit.

where c, e and ϵ are the effective elastic, piezoelectric and dielectric constants for the particular excited mode, and are obtained from the defining relations Eqns (2.12), (2.13) and (2.16) when these are expressed in the coordinate system that diagonalizes L_{km}.

Although the expressions above are derived only for the special case of pure thickness vibrations, and so cannot be expected to yield accurate numerical results, they nevertheless demonstrate important functional dependencies of the equivalent circuit parameters that are borne out in practice. In particular, the reduction in C_1 and in the capacitance ratio C_1/C_0 with the square of the overtone order means that the 'pulling sensitivity' (Chapter 6) of overtone units is at least an order of magnitude less than that for fundamentals.

In all the analysis of this chapter it has been assumed that the material is non-dissipative, and consequently the derived equivalent circuit contains no resistive element. This assumption can be removed by assuming that the stresses depend not only on the strain but also on the time rate of change of the strain. Ignoring the piezoelectricity for the moment, this means that the stress-strain relationships have to be generalized to

$$T = cS + \eta \dot{S}$$

where the tensor indices have been dropped for simplicity, but nevertheless the *viscosity* η is a fourth rank tensor having the same symmetry properties as the elastic constants c. In a situation where the time dependence is of the form $\exp(j\omega t)$, $\dot{S}$ is just $j\omega S$, and thus the stress-strain relationship is

$$T = (c + j\omega\eta)S$$

In this case then the viscous losses can be taken into account by treating the elastic constants as complex rather than real quantities. The analysis of Section 2.1 can be carried forward with complex elastic constants just as before, up to and including the derivation of the wave equation (Eqn 2.15). The tensor L_{km} will now have complex components through the definitions in Eqns (2.11) and (2.16) and will therefore have complex eigenvalues c_m. However, provided the losses are very small, the complex eigenvalue problem can be avoided by ignoring the imaginary part of L_{km} and diagonalizing the real part. The imaginary part of the eigenvalue c_m, that is, the effective viscosity in the mth mode, can then be restored by using the analogue of Eqn (2.11), so that

$$\delta_{km}\eta_m = \eta_{klmn}n_l n_m \tag{2.52}$$

with all quantities expressed in the coordinate system in which L_{km} is diagonal.

In the wave equation (2.19), c_k is then to be replaced by the complex $c_k + j\omega\eta_k$. The assumed solutions given in Eqn (2.32) must then be augmented by an additional factor $\exp(-\alpha t)$, where α is given by

$$\alpha = (\eta_k \omega^2)/(2c_k) \tag{2.53}$$

Comparing α as given by Eqn (2.53) to the corresponding expression for the damping factor in a series RLC circuit with a quality factor Q leads to the identification

$$R/2L = \omega/(2Q) = (\eta_k \omega^2)/(2c_k)$$

and thus to the following expression for the intrinsic Q of the kth mode

$$Q_k = c_k/(\omega \eta_k) \tag{2.54}$$

In terms of the *time constant* τ_k introduced by Guttwein, Lukaszek and Ballato (Guttwein *et al.*, 1967) and defined by $\tau_k = \eta_k/c_k$

$$Q_k = 1/\omega \tau_k \tag{2.55}$$

If a resistance R_1 is included in the series arm of the equivalent circuit of Fig. 2.2, then the Q of the series arm becomes $Q = \omega L_1/R_1$ or $Q = 1/(\omega R_1 C_1)$. Hence $\tau_k = R_1 C_1$ and so from Eqn (2.51)

$$R_1 = M^2 \pi^2 \eta_k h/(4e^2 A) \tag{2.56}$$

The equivalent circuit including R_1 is shown in Fig. 2.3. The expression (Eqn 2.56) for R_1 only includes the effect of the acoustic attenuation due to the viscosity η of the standing wave system in the resonator. In an actual resonator there will be additional contributions to R_1 due to such factors as mounting losses, atmospheric damping, losses in the electrodes and losses due to imperfect surface finish on the major surface of the resonator. Hence the preceding expression for R_1 should be regarded as giving a lower bound for the R_1 of practical resonators, albeit one which can be approached quite closely by proper resonator design and careful processing.

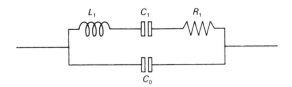

Fig. 2.3 Equivalent circuit of crystal resonator.

2.5 TEMPERATURE COEFFICIENTS OF FREQUENCY

Equation (2.42) determines the anti-resonance frequencies of thickness modes in an infinite plate in terms of the effective elastic constant c_m for the mth mode, the density ρ, and the plate thickness $t = 2h$. The elastic constant

c_m is in turn determined in terms of the fundamental elastic, piezoelectric and dielectric constants through the solution of the eigenvalue problem for the symmetric tensor $\bar{L}_{km}$ defined in Eqn (2.16).

Since all of the material constants and the plate thickness vary with temperature, so too will the anti-resonance frequencies. For a given reference temperature T_0, the rth order temperature coefficient of a function f of the temperature T is defined by

$$T_f^{(r)} = f^{(r)}(T_0)/(r!\, f(T_0)) \tag{2.57}$$

where $f^{(r)}(T_0)$ denotes the rth derivative of $f(T)$ with respect to T evaluated at T_0. By repeatedly differentiating Eqn (2.42) with respect to T, the temperature coefficients of the frequency can be expressed in terms of the coefficients of the effective elastic constant, the density and the plate thickness. In particular, the first-order coefficient of frequency can be immediately obtained as

$$T_f^{(1)} = T_{c_m}^{(1)}/2 - T_\rho^{(1)}/2 - T_t^{(1)} \tag{2.58}$$

$T_\rho^{(1)}$ is independent of the orientation of the plate and is completely determined by the thermal expansion properties of the material. $T_t^{(1)}$ is the effective coefficient of linear expansion in the direction of the plate normal n_k, and thus depends on the thermal expansion coefficients and the orientation. In the notation of Chapter 1, if S_{km} is the thermally induced strain at a temperature T referred to the reference state at temperature T_0, and if $\alpha_{km}^{(r)}$ are the expansion coefficients of order r, then

$$S_{km}(T) = \alpha_{km}^{(1)}(T - T_0) + \alpha_{km}^{(2)}(T - T_0)^2 + \alpha_{km}^{(3)}(T - T_0)^3$$

or to first order

$$S_{km}(T) = \alpha_{km}^{(1)}(T - T_0)$$

From the definition (Appendix 3) of the strain tensor, the fractional change in volume of an elementary volume element accompanying a strain S_{km} is just $S_{kk} = S_{11} + S_{22} + S_{33}$, so that since mass is conserved, the corresponding fractional change in the density is just $-S_{kk}$. Hence the first-order temperature coefficient of the density is just

$$T_\rho^{(1)} = -\alpha_{kk}^{(1)} = -(\alpha_{11}^{(1)} + \alpha_{22}^{(1)} + \alpha_{33}^{(1)}) \tag{2.59}$$

Also, again from the definition of S_{km}, the fractional change in length of a line element along the direction n_k is $S_{km}n_k n_m$, so that the effective coefficient of linear expansion along n_k is

$$T_t^{(1)} = \alpha_{km}^{(1)} n_k n_m \tag{2.60}$$

From the data given in Table 1.1, it follows that $T_\rho^{(1)}$ has the value -34.9 ppm/°C. The maximum value for $T_t^{(1)}$ occurs when n_k is perpendicular to the optic axis and is $\alpha_{11}^{(1)}$ or 13.71 ppm/°C, while the minimum occurs when n_k is along the optic axis and is $\alpha_{33}^{(1)}$ or 7.48 ppm/°C. Hence the contribution of the last two terms in Eqn (2.58) to the first-order temperature coefficient of frequency varies from a minimum of +3.74 ppm/°C with n_k perpendicular to the optic axis, to a maximum of +9.97 ppm/°C for n_k along the optic axis. Since the values of the first-order coefficients of the fundamental elastic constants, which are the major determinants of the effective elastic constant c_m, are typically an order of magnitude greater than these contributions, it follows that the temperature coefficient of frequency is primarily determined by the temperature coefficients of the elastic constants.

So far, the discussion has been limited to the anti-resonance frequencies, although in practice it is usually the resonance frequencies that are of more interest. As has already been pointed out in Section 2.3, in the general case where all three thickness modes are excited, the resonance frequencies have to be determined as solutions of Eqn (2.43). Fortunately, for the important quartz cuts it is usually sufficient to consider the case where either only one mode is excited, or where, even if two or more modes are present, their resonance frequencies are sufficiently far separated for the single-mode analysis to be used.

The key result in the single-mode case is Eqn (2.48), which relates the fractional frequency difference between the anti-resonance and resonance frequencies to the electromechanical coupling factor for the mode under consideration. Writing f_A and f_R for the anti-resonance and resonance frequencies and dropping the suffix m identifying the particular mode, Eqn (2.48) can be rewritten

$$(f_A - f_R)/f_A = 4k^2/M^2\pi^2$$

or

$$f_R = f_A (1 - 4k^2/M^2\pi^2) \tag{2.61}$$

Differentiating Eqn (2.61) and making use of the fact that in quartz the coupling factor k^2 is always $\ll 1$ leads then to the result

$$T_{f_R}^{(1)} = T_{f_A}^{(1)} - 8k^2 \, T_k^{(1)}/M^2\pi^2 \tag{2.62}$$

with $T_k^{(1)}$ being the first-order temperature coefficient of the coupling factor. It follows immediately from Eqn (2.62) that the difference between the temperature coefficients of the resonance and anti-resonance frequencies decreases rapidly with both increasing overtone order and decreasing coupling.

Analytical expressions for the higher order temperature coefficients of frequency in terms of the higher order coefficients of the material constants can be developed by further differentiation of the basic frequency equation (2.42). However, these expressions rapidly become cumbersome as the order increases, and it is usually easier to determine higher order coefficients by numerical methods.

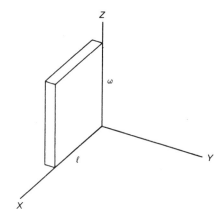

Fig. 2.4 (YX) crystal plate.

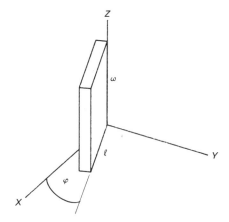

Fig. 2.5 (YXw) ϕ crystal plate.

2.6 DOUBLY ROTATED CUTS

The 1978 *IEEE Standard on Piezoelectricity* (IEEE, 1978) describes a general method of specifying a crystal plate that is arbitrarily oriented with respect to the crystallographic axes $OXYZ$. The application of this method to generally oriented thickness mode resonators is as follows. Choose as a starting point a Y-cut plate of length l, width w and thickness t, with the length along OX, width along OZ and thickness along OY (Fig. 2.4). This plate has the simple notation (YX), where the first symbol indicates the axis along which the thickness lies and the second the axis along which the length l lies. The (YX) plate is then rotated about its width, that is about OZ, through an angle ϕ (Fig. 2.5). This singly rotated plate then has the notation (YXw)ϕ

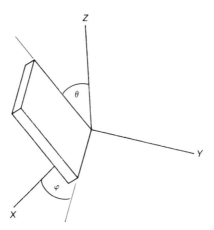

Fig. 2.6 (*YXwl*) ϕ/θ crystal plate.

where w indicates the axis of the rotation and ϕ its magnitude. A second rotation of magnitude θ is then applied about the length of the (*YXw*) plate, to produce a *doubly rotated* plate described by the notation (*YXwl*) ϕ/θ (Fig. 2.6).

The IEEE standard goes on to describe triply rotated plates, but for pure thickness modes the doubly rotated plate covers all possibilities as the third rotation would be about the plate normal and would therefore not alter the physical situation. (This of course assumes a plate of effectively infinite lateral extent.) In the case of quartz, having trigonal symmery about the optic axis OZ and digonal symmetry about OX, it is easily seen that for a doubly rotated plate, the angle range $\phi = 0°$ to $\phi = 30°$ covers all eventualities, with θ ranging over $-90°$ to $+90°$. The plate normal n_k initially lies along OY and so has components $(0,1,0)$. The components n_k of the normal to the doubly rotated plate are obtained by applying the rotation matrices (Appendix 5) for the ϕ and θ rotations to the initial vector $(0,1,0)$, and are

$$\begin{aligned} n_1 &= -\cos(\theta)\sin(\phi) \\ n_2 &= +\cos(\theta)\cos(\phi) \\ n_3 &= \sin(\theta) \end{aligned} \qquad (2.63)$$

The analysis of thickness modes in the preceding sections, together with a set of values for the fundamental material constants and their temperature coefficients, in principle allows the theoretical prediction of the key properties of a thickness mode resonator as functions of the orientation angles ϕ and θ. In particular, the frequency–thickness constants and the electromechanical coupling factors for each mode can be calculated as a function of temperature and orientation. By curve-fitting the data so obtained, the first and higher order temperature coefficients of the anti-

Table 2.1 Doubly rotated crystal cuts

Cut	AT	FC	IT	SC	LC	BT	RT
ϕ	0	15	19.10	21.93	11.17	0	15
θ	−35.25	−34.33	−34.08	−33.93	−9.39	49.20	34.50
k_a	0	2.37	2.96	3.33	3.21	0	4.27
k_b	0	3.61	4.33	4.71	7.64	5.62	6.46
k_c	8.8	6.89	5.79	4.99	9.21	0	2.12
N_a	3504	3446	3411	3382	3165	3089	3059
N_b	1900	1936	1959	1977	2140	2536	2260
N_c	1661	1726	1766	1797	1727	1884	2040
$T_a^{(1)}$	−48.9	−50.1	−51.2	−52.1	−26.1	−95.6	−74.6
$T_b^{(1)}$	−31.3	−29.1	−27.5	−26.2	−39.7	0	−1.49
$T_c^{(1)}$	0	0	0	0	39.8	−30.9	0

Note: 1 Coupling factors k_a, k_b and k_c in %.
2 Frequency constants N_a, N_b and N_c in kHz mm.
3 Temperature coefficients $T_a^{(1)}$, $T_b^{(1)}$ and $T_c^{(1)}$ in ppm/°C.
For further detail, refer to Ballato (1977).

resonance frequencies, the coupling, and the resonance frequencies can easily be obtained.

Table 2.1 lists some of these principal characteristics of the better known doubly rotated cuts. The Table gives the commonly used designation for each cut, the approximate ϕ and θ angles, and the coupling factors, frequency constants, and first-order frequency–temperature coefficients for the three possible thickness modes in each orientation. The modes are identified in accordance with the usual conventions (Ballato, 1977) as a, b and c modes, with the phase velocities being ordered as $V_a > V_b > V_c$. Although the classification of the modes as longitudinal and transverse is not generally valid (cf. Section 2.1), nevertheless the a mode is commonly referred to as the *quasi-longitudinal* mode, and the b and c modes as the *fast shear* and *slow shear* mode, respectively. (It should be noted that in Table 2.1 and elsewhere in this book, the 1978 *IEEE Standard on Piezoelectricity* has been followed in adopting sign conventions for the crystallographic axes. This means that the sign of the θ angle is reversed compared to the conventions adopted in much of the earlier work.)

The special cases of the AT- and BT-cuts, which have $\phi = 0°$ and belong to the rotated Y-cut family, are considered in more detail in the following sections, but are included here for purposes of comparison. All of the cuts listed with the exception of the LC-cut have a zero first-order temperature coefficient of frequency for either the b mode or the c mode at around room temperature. The LC-cut is included because the second and third order coefficients for the c mode vanish, resulting in an ultra-linear frequency–temperature characteristic useful in precision thermometry.

The IT cut (Bottom and Ives, 1951), RT-cut (Bechmann, 1961) and the FC-cut (Lagasse et al., 1972) are all cuts introduced for the main purposes of improving upon the well established AT-cut's performance in certain applications. The typical AT frequency–temperature characteristic (Section 2.8) is cubic in character, with a negative slope at room temperature and turning points symmetrically located about a point of inflexion at or just above room temperature. In applications where the frequency stability requirements are such as to make temperature control of the crystal necessary, it is standard practice to place the crystal in an oven whose operating temperature is set to the upper turnover point of the crystal, so that the frequency–temperature coefficient is zero at the oven temperature. For the AT this requires that oven temperature and crystal turnover be closely matched in order to obtain acceptable results.

The principal difference between the AT- and the FC-cut is that in the latter the inflexion temperature is much higher. This means that for a given oven temperature T, an FC-cut with an upper turning point T will have a relatively much flatter frequency–temperature curve than a corresponding AT. This can be taken advantage of in one of two ways: either to obtain improved stability from a given oven, or to relax the oven requirements in respect of temperature stability while maintaining the frequency stability. A subsidiary advantage of the higher inflexion point is that the frequency deviation from room temperature to oven temperature is much less than with the AT. These points are illustrated in Fig. 2.7.

Although of considerable significance, these advantages have not been the major factor in the increased use of doubly rotated cuts. The predictions in 1974 and 1975 of plate orientations minimizing certain non-linear effects, along with their later experimental confirmation, provided the impetus for the continuing use of doubly rotated resonators. Holland (1974a, b) first predicted that at an orientation of $\phi = 22.8°$, $\theta = -34.3°$ thermal shocks would produce no frequency excursions or transients, and coined the term

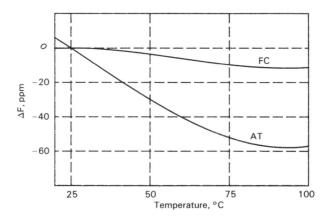

Fig. 2.7 Comparison of AT and FC characteristics.

TS-cut (for *t*hermal *s*hock). In the following year, EerNisse (1975) similarly predicted that at an orientation of $\phi = 22.5°$, $\theta = -34.3°$, frequency changes due to mechanical stresses in the plane of the resonator plate would be minimized and introduced the term SC-cut for *s*tress *c*ompensated). It is now generally recognized that these cuts are essentially the same, and the SC-cut nomenclature is commonly accepted. The main advantages and disadvantages of the SC-cut versus the AT-cut have already been summarized in Section 1.4.7.

2.7 ROTATED Y-CUTS

In an arbitrarily oriented crystal plate, the effective symmetry of the device is much reduced as compared to the symmetry of the material. In general both the electric and the optic axes, which are, respectively, digonal and trigonal symmetry axes, will make oblique angles with the edges of the plate. However, in those special cases where the plate axes include one of the crystal symmetry axes, it is to be expected that the performance of the resonator will reflect this symmetry in some respects. This is true in particular of the rotated Y-cut family of plates $(YXwl)\phi/\theta$ with $\phi = 0°$. Since $\phi = 0$, an equivalent notation is simply $(YXl)\ \theta$.

In such plates, the length l is originally along the X or electric axis, and this is the axis about which the plate is rotated. Thus the X axis is contained in the plane of the plate for all members of the rotated Y-cut family, and consequently the plate normal n_k has no component along X, $n_1 = 0$. From Eqn (2.63), the components of n_k are just $(0, \cos(\theta), \sin(\theta))$.

Setting $n_1 = 0$ and using the symmetry relations that exist among the fundamental material constants of quartz (Section 1.3), the effective dielectric and piezoelectric constants ϵ and e_k defined in Eqns (2.13) and (2.12) reduce to the simple forms

$$\epsilon = \epsilon_{km} n_k n_m = \epsilon_{11} n_2^2 + \epsilon_{33} n_3^2 \tag{2.64}$$

$$\begin{aligned} e_1 &= -e_{11} n_2^2 - e_{14} n_2 n_3 \\ e_2 &= e_3 = 0 \end{aligned} \tag{2.65}$$

Similarly, from the defining equation for L_{km}, Eqn (2.11),

$$\begin{aligned} L_{11} &= c_{66} n_2^2 + c_{44} n_3^2 + 2 c_{14} n_2 n_3 \\ L_{12} &= L_{13} = 0 \end{aligned} \tag{2.66}$$

Using Eqns (2.65) and (2.66) with the definition Eqn (2.16) of $\bar{L}_{km}$ gives finally

$$\begin{aligned} \bar{L}_{11} &= L_{11} + e_1^2/\epsilon \\ \bar{L}_{12} &= \bar{L}_{13} = 0 \end{aligned} \tag{2.67}$$

Equations (2.65) and (2.67), in conjunction with the wave equation (2.15), the expression for the electrostatic potential (2.14), and the expressions for the surface tractions and electric displacement in Eqns (2.17) and (2.18), show that the mechanical displacement u_1 is not coupled to the displacements u_2 and u_3, and is the only displacement coupled to the electric field. This leads directly to the key result that for the rotated Y-cuts, the only one of the three possible thickness modes that can be piezoelectrically excited must be a pure thickness shear with the displacement along the X axis. The relevant wave equation is

$$\bar{L}_{11} u_{1,xx} = \rho \ddot{u}_1 \tag{2.68}$$

The accompanying electric potential is

$$\phi = e_1 u_1 / \epsilon + Cx + D \tag{2.69}$$

and the electromechanical coupling factor is

$$k^2 = e_1^2 / \epsilon \bar{L}_{11} \tag{2.70}$$

From Eqn (2.42), the frequency–thickness constant N follows as

$$N = 2hf = (\bar{L}_{11}/\rho)^{1/2}/2 \tag{2.71}$$

Figures 2.8 and 2.9 show the variation of the coupling factor and the frequency–thickness constant with the rotation angle θ. Because the term e_1^2/ϵ in Eqn (2.67) represents only a small correction to L_{11}, to a good approximation the extreme values of N coincide with those of L_{11}. Differentiation of Eqn (2.66) shows that these extrema occur at angles such that

$$\tan(2\theta) = 2c_{14}/(c_{66} - c_{44}) \tag{2.72}$$

Reference to Appendix 5 shows that this also gives the angles at which the elastic constant c_{56}', expressed in the plate coordinate system, vanishes. As

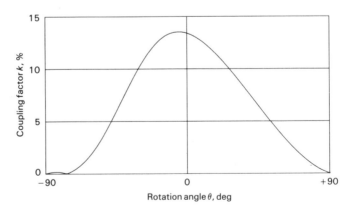

Fig. 2.8 Coupling factor for rotated Y-cut plates.

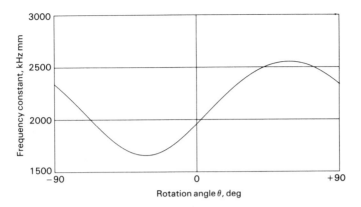

Fig. 2.9 Frequency constant for rotated Y-cut plates.

already stated in Section 1.4.7, the solutions of Eqn (2.72) are $-31.6°$ and $+58.4°$, corresponding, respectively, to the AC- and BC-cuts.

The first-order temperature coefficient of the anti-resonance frequency is determined by Eqn (2.58), with the dominant term being $T^{(1)}_{c_m}$. In the present case, the effective elastic constant is $\bar{L}_{11}$, or to a good approximation, L_{11}. Differentiating Eqn (2.66) with respect to the temperature leads to the following expression for the first-order coefficient of L_{11} in terms of the coefficients of the fundamental elastic constants

$$T^{(1)}_{L_{11}} = (c_{66}T^{(1)}_{c_{66}}n_2^2 + c_{44}T^{(1)}_{c_{44}}n_3^2 + 2c_{14}T^{(1)}_{c_{14}}n_2n_3)/L_{11} \tag{2.73}$$

Fig. 2.10 shows $T^{(1)}_{L_{11}}$ as a function of orientation. There are two angles θ where small negative values of $T^{(1)}_{L_{11}}$ compensate for the net positive contribution of $T^{(1)}_\rho$ and $T^{(1)}_t$ to give a zero temperature coefficient of frequency. These angles correspond to the AT- and BT-cuts, at $\theta = -35.25°$ and $\theta = +49.2°$, respectively.

2.8 AT-CUT RESONATORS

Table 2.2 lists the values of the key physical parameters for the AT-cut resonator as determined from the expressions derived in the previous section. The term AT-cut is in practice applied to resonators that have a range of θ values of the order of 1° around the nominal AT angle, and there will be slight variations in the values of the parameters in Table 2.2 depending on the precise angle used. However, the errors resulting from the thickness mode approximation generally make it unnecessary to take account of errors due to slight angle differences.

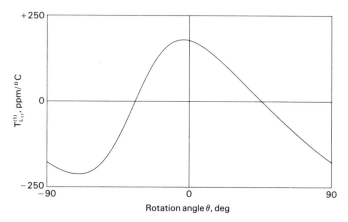

Fig. 2.10 $T_{L_{11}}^{(1)}$ for rotated Y-cut plates.

Table 2.2 Physical parameters for AT-cut resonators

ϕ angle	0°
θ angle	−35.25°
Density	2649 kg/m³
Effective elastic constant	29×10^9 N/m²
Effective piezoelectric constant	0.095 C/m²
Effective dielectric constant	40.3×10^{-12} F/m
Frequency constant	1660 kHz/mm
Coupling factor	8.74%

The defining feature of both AT- and BT-cuts is the vanishing of their first-order temperature coefficients at room temperature. In principle, the higher order coefficients could be calculated from the corresponding coefficients of the fundamental material constants, but in practice it turns out that the frequency–temperature characteristics of both AT- and BT-cuts are known empirically with more precision than can be obtained by calculation. Thus the situation is often reversed, in that the material properties are inferred from the frequency–temperature characteristics rather than vice versa.

Bechmann (1955, 1956, 1960) has established that for both AT- and BT-resonators, their frequency–temperature characteristics over a wide temperature range can be adequately described by a power series in the temperature including terms up to third order. Writing f for the frequency at a temperature T, and f_0 for the frequency at a reference temperature T_0, Bechmann's expression is

$$(f - f_0)/f_0 = a(T - T_0) + b(T - T_0)^2 + c(T - T_0)^3 \tag{2.74}$$

Bechmann also found that within small angle ranges about the nominal angles for the AT and BT, the coefficients a, b and c could be expressed as linear functions of the angle increment $\Delta\theta$ from some predefined reference angle. Thus a can be written

$$a = a_0 + a_1\Delta\theta \tag{2.75}$$

with similar expressions for b and c.

Although Eqn (2.74) applies to both AT and BT resonators, the relative magnitude of the various terms is quite different. In the AT the cubic term is the dominant one, giving rise to the typical 'S' shaped characteristics shown in Fig. 2.11, whereas the quadratic term is dominant in the BT. Hence the BT characteristics are parabolic in nature, as shown in Fig. 2.12, making the BT much less suitable for applications involving wide temperature ranges.

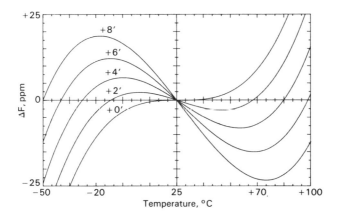

Fig. 2.11 Normalized AT-cut curves.

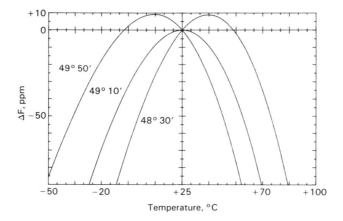

Fig. 2.12 Normalized BT-cut curves.

Table 2.3 Bechmann's coefficients for the AT-cut resonator

Reference temperature	20°C
Reference angle	−35°15'
Coefficients	
a_0	0
a_1	$-5.15 \times 10^{-6}/°C/°\theta$
b_0	$0.39 \times 10^{-9}/(°C)^2$
b_1	$-4.7 \times 10^{-9}/(°C)^2/°\theta$
c_0	$109.5 \times 10^{-12}/(°C)^3$
c_1	$-2.0 \times 10^{-12}/(°C)^3/°\theta$

Table 2.3 gives Bechmann's 1960 values of a_0, a_1, etc, for the AT resonator, referred to $T_0 = 20°C$ and a reference angle of $-35.25°$. It should be borne in mind in using this data that the values were arrived at by analysis of results from a wide range of resonators of different detailed design, and should be applied with caution in particular cases. In particular, the choice of reference angle required to fit the measured data in a specific case is a sensitive function of the resonator design, depending strongly on the overtone, the geometry of the blank, and the electrical operating conditions. These are considered in more detail in Chapter 7.

Nonetheless, the main features of the frequency–temperature characteristics of AT cut resonators can be seen using the data of Table 2.3. Fig. 2.11 shows the typical characteristics for a range of angle increments $\Delta\theta$. As the angle increases, the slope of the curves at room temperature becomes increasingly negative, while at the same time the separation between the turning points and the accompanying frequency deviation increases. The large scope for trading frequency stability against operating temperature range is apparent from the Figure, and has been the main factor in making the AT resonator the most commonly used crystal unit in a wide range of applications.

3 Energy trapping

Within the linear theory of piezoelectric materials, simple and exact solutions are available for the idealized problem of thickness modes in thin plates. These solutions are invaluable for the basic understanding they give of the physical situation, but are not by themselves sufficient to give either a full appreciation of observed resonator behaviour, or to predict precise numerical results. Extensions to the simple theory are required in two directions. Firstly, non-linear effects are explicitly excluded from consideration in what is by definition a linear theory. Secondly, and most immediately important, the restriction to pure thickness modes has to be removed to give an understanding of such key phenomena as *energy trapping* and the coupling of the thickness modes to other modes dependent on the finite lateral dimensions of the plate.

3.1 WAVE PROPAGATION IN THIN PLATES

Pure thickness modes have a straightforward physical interpretation in terms of standing wave patterns set up by the reflection at the major surfaces of the plate of plane waves travelling in the thickness direction. The assumption of plane waves implies the assumption that the lateral dimensions of the plate, that is the length and width of a rectangular plate or the diameter of a circular plate, can be regarded as infinite. In practice, the ratio of the lateral dimensions to the thickness can vary from several hundred to one for very high frequency plates, to less than ten to one for low frequencies. Consequently, the thickness mode analysis cannot be expected to be anything more than a rough guide to the behaviour of plates in the low frequency region.

To take account of the finite lateral dimensions of an actual resonator, the most natural approach is to extend the interpretation in terms of standing wave patterns. That is, as well as considering waves propagating in the thickness direction and being reflected from the major surfaces, waves should also be considered that propagate in the plane of the plate and are reflected at the plate edges. An analysis of this sort would then be expected to cover not only the influence of the finite lateral dimensions on the thickness modes, but also as special cases the length extensional, flexural and face shear modes discussed in Chapter 1.

A prerequisite of such a program is a clear understanding of the propagation of waves in unbounded plates. As shown in the previous chapter, in an unbounded material, for any given direction of propagation, there are just three possible types of plane wave solution, making analysis straightforward. Additionally, the phase velocities of the waves are constants, independent of wave number, so that if the frequency ω is plotted against wavenumber k, the result is the set of three straight lines shown in Fig. 3.1. The slopes of the lines are just the three characteristic phase velocities V_k (Section 2.1). In this case, the group velocity $d\omega/dk$ for each mode is also constant and equal to the phase velocity. In general, a plot of ω versus k for a given type of wave or mode is termed the dispersion relation for that mode. The mode is termed dispersive when the phase velocity depends on the wavenumber, that is when the group velocity $d\omega/dk$ is not constant. In such case, the phase and group velocities are no longer equal.

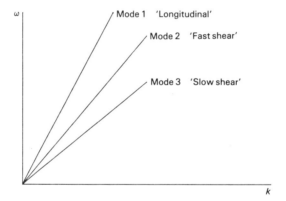

Fig. 3.1 Dispersion relations for plane waves.

In the case of waves in the plane of an unbounded plate, say with major surfaces at $y = x_2 = \pm h$, the situation is vastly more complicated as compared to the thickness mode case. The essential physical model of plate waves as being made up of combinations of plane waves successively reflected from the major surfaces remains correct, but it turns out that the number of possible combinations gives rise to many different types of plate wave. These can be broadly classified according to the nature of the particle displacements involved as *flexural* (F), *extensional* (E), *face shear* (FS), *thickness shear* (TS) and *thickness twist* (TT). Figure 3.2 illustrates the displacements for each type of wave.

The essential analysis required to obtain the dispersion curves for thickness

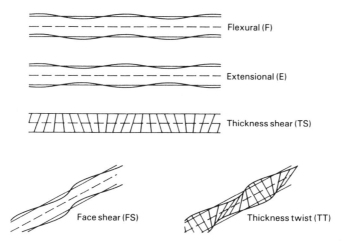

Fig. 3.2 Principal displacements for plate waves.

shear, flexural and extensional waves in *isotropic* plates was first given by Rayleigh (1889). As shown in Appendix 6, the satisfaction of the boundary conditions for stress-free major surfaces results in a transcendental equation involving the wavenumbers in both the thickness direction and the direction of propagation. The solutions of this equation determine the various branches of the dispersion relation for the isotropic plate. The dispersion curves for thickness twist and face shear modes are also obtained in Appendix 6 for the isotropic case, the analysis for these modes being very much simpler than for the thickness shear group. Figures 3.3 and 3.4 show very much simplified sketches of the normalized dispersion curves for those modes of most relevance in resonator theory, including only face shear, flexural,

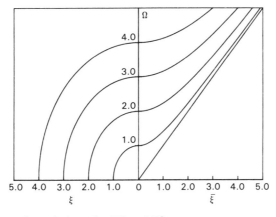

Fig. 3.3 Dispersion relations for TT and FS waves.

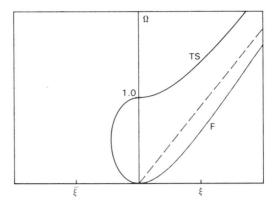

Fig. 3.4 Dispersion relations for TS and F waves.

thickness shear and thickness twist modes. The frequency Ω is normalized with respect to the lowest thickness mode frequency, whereas the wavenumber along the plate ξ is expressed in terms of the number of half wavelengths that would be contained in the thickness of the plate. In the Figures, the right-hand portion of the curves refers to real wavenumbers ξ, and the left-hand portion refers to imaginary ξ. The physical significance of this is that for imaginary wavenumbers, the particular mode in question is no longer a travelling wave, but an *evanescent* wave whose amplitude decreases exponentially with distance, and which is incapable of transmitting energy.

The important feature of the curves in Figs. 3.3 and 3.4 is that for both TT and TS modes, cut-off frequencies exist such that for frequencies below the cut-off for a particular mode, that mode exists only as an evanescent wave. As shown in Appendix 6, the cut-off frequencies correspond to the thickness mode frequencies. A simple physical explanation can be given of the cut-off phenomenon in terms of the model of the plate wave based on successive reflections of plane waves from the plate surfaces. In this model, there are clearly two distinct limiting situations. In the first, the component plane waves are actually travelling along the plate, parallel to the major surfaces, so that no reflections occur. Then the plate wave velocity coincides with the bulk wave velocity. As the angle of incidence of the component plane waves decreases from 90° towards normal incidence, then it is physically clear that the rate of transfer of energy *along* the plate, that is the *group velocity*, must decrease because of the increased path length due to an increased number of reflections. Clearly, at normal incidence, the second limiting case is reached, which is just that of pure thickness vibrations, with no energy being transmitted along the plate. This is the cut-off condition, in which the group velocity of the plate wave is zero.

3.2 RESONANCE FREQUENCIES FOR FINITE PLATES

Given a knowledge of the characteristics of waves in unbounded plates, the problem of determining the normal modes of vibration of a finite plate can in principle be regarded as equivalent to the problem of determining a linear combination of plate waves that will satisfy the boundary conditions on the edges of the plate. In this approach, since all the plate waves individually satisfy the boundary conditions on the major surfaces, it is only the edge conditions that need to be taken into account. Unfortunately, in all but the simplest limiting cases, solution in terms of a finite number of wave functions do not exist. This is so even in the case of isotropic plates; the additional complications of anisotropy and piezoelectricity only serve to complicate matters.

In the absence of exact solutions to the general problem, a great deal of work has been done in determining approximate solutions. There have been essentially two distinct approaches. The first uses the full set of field equations and boundary conditions and seeks to determine approximate solutions by such methods as perturbation theory or variational techniques. This approach is treated in detail by Holland and EerNisse (1969). The second approach is to use whatever special features of the problem exist to reduce the full set of equations to a more manageable, approximate set that can be solved exactly. In the context of resonator theory, the pioneering work in this area has been done by Mindlin and his co-workers in a series of papers beginning in 1951. The main features of this work are described by Tiersten (1969).

For present purposes, the details of both approaches can be passed over. The essential physical principles can be understood in simple terms by adopting as a general rule, applicable to any type of resonating system where a particular type of wave motion is predominant, the assumption that the condition for resonance is that an integral number of half wavelengths be contained in the dimension along the direction of propagation.

For thickness modes in rotated Y-cut quartz plates, it has been shown in Chapter 2 that the only piezoelectrically excited mode has $u = u_1$ as the only non-zero displacement. Consequently, in considering the modes of vibration of finite plates it is reasonable to assume that u will continue to be the major displacement component. From the discussion of isotropic plates in Appendix 6, the plate waves of immediate interest will then be thickness twist and face shear modes propagating in the $z = x_3$ direction, and thickness shear and flexural modes propagating in the x direction. In other words, the model of the rotated Y-cut resonator in terms of pure thickness modes is to be generalized by considering these specific types of wave motion in the plate.

For example, consider a plate with major faces at $y = \pm h$ as before, unbounded in the z direction, but with edges at $x = \pm L$. Then for those resonances associated with flexural waves propagating along x, the above

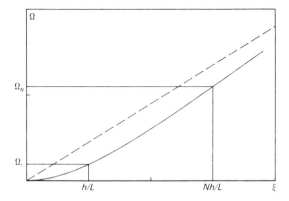

Fig. 3.5 Resonance frequencies for flexural modes.

rule, in conjunction with the dispersion relation for flexural waves, allows estimates to be made of the resonance frequencies as follows. First, from the definition of the normalized wavenumber ξ, the ξ values corresponding to the resonance condition are just

$$\xi = Nh/L \tag{3.1}$$

Figure 3.5 shows the dispersion relation for flexural modes, with the points Nh/L marked off along the ξ axis. The normalized resonance frequencies Ω_N are then determined as the frequency values corresponding to these ξ values. Notice that for Ω_N to be comparable with $\Omega = 1$, the lowest thickness shear frequency, N must be of the order of L/h, the length-to-thickness ratio of the plate.

The same argument may be applied to the estimation of the thickness shear frequencies of the plate. Figure 3.6 shows the dispersion relation for the TS

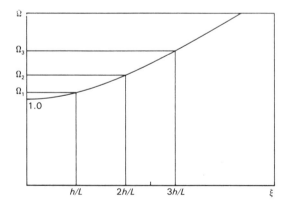

Fig. 3.6 Resonance frequencies for TS modes.

branch, again with the points Nh/L marked off on the ξ axis. Once again there are a whole family of resonances associated with the TS waves, but because of the shape of the dispersion curves the resonance frequencies are clustered just above the cut-off frequency. The lowest frequency mode is the main resonance associated with that particular branch of the dispersion relation, whereas the remaining modes are termed the *inharmonic modes* for that branch. For large L/h values, the frequency of the lowest or main resonance tends towards the cut-off frequency, that is the pure thickness mode frequency, just as would be expected.

For a strip resonator of infinite extent in the x direction but bounded by edges at $z = \pm W$, an exactly analogous discussion applies with face shear and thickness twist waves playing the role of flexural and thickness shear waves. Thus there will be a family of inharmonic thickness twist modes and a family of face shear resonances. In the more general case of a rectangular plate bounded in both directions, then the expectation must be that the inharmonic responses will consist of both thickness twist and thickness shear modes, and that both flexural and face shear resonances will be found. A more precise analysis reveals that not only is this the case, but that more complicated resonances exist, such as thickness twist overtones of the inharmonic thickness shear modes (Mindlin and Spencer, 1967). Nevertheless, the essential physical picture of the finite plate remains as one where the pure thickness mode suffers a slight upward displacement in frequency and is accompanied by a cluster of closely spaced inharmonics at higher frequencies, together with separate families of flexural and face shear modes.

At this point it is convenient to note that the inharmonics, as resonances associated with the same branch of the dispersion relation as the main thickness response, are governed by the same set of material constants. Consequently, their frequency–temperature characteristics can be expected to track the characteristic of the main response. However, the flexural and face shear modes, being associated with different branches of the dispersion relation and thus with a different set of material constants, can generally be expected to show quite different temperature characteristics. In practice, for the AT- and BT-cuts, these modes generally have large negative temperature coefficients compared to an essentially zero coefficient for the thickness mode and its inharmonics.

3.3 COUPLED MODES IN FINITE RESONATORS

Figure 3.7 shows again the dispersion relations for the flexural and thickness shear modes in a plate with edges at $x = \pm L$. Following the previous discussion the main thickness mode will have the resonance frequency Ω determined by the TS branch at the wavenumber $\xi = h/L$. At this frequency,

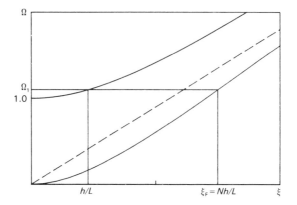

Fig. 3.7 Condition for coupling between TS and F modes.

the flexural mode will have the wavenumber ξ_F as shown in Fig. 3.7. If it should happen that $\xi_F = Nh/L$ for some integer N, then clearly the resonance frequencies of the main thickness mode and the Nth flexural mode coincide. In such a case, the basic assumption underlying the previous discussion is no longer valid, since it is no longer true that the resonance is primarily associated with a single type of wave motion.

Figure 3.8 shows in schematic form how the thickness modes of a strip resonator with edges at $x = \pm L$ vary with the length to thickness ratio L/h. As shown already, as L/h increases the frequencies tend in the limit to the frequencies of the pure thickness modes, and for reasonably large L/h values the variation is small. Also shown in the figure is the dependence of the frequency of a flexural mode with large N on the parameter L/h. Examination of the dispersion curves for flexural modes shows that for large wavenumbers, the dispersion relation approximates a straight line. This in

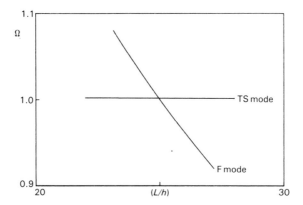

Fig. 3.8 TS and F mode frequencies vs (L/h).

turn implies that in this region the propagation of flexural waves is non-dispersive and that the dispersion curve can be approximated by a linear expression

$$\Omega = V\xi/V_s \tag{3.2}$$

where V is the phase velocity. Since for the Nth flexural mode in a strip, $\xi = Nh/L$, it follows that

$$\Omega = N(V/V_s)/(L/h) \tag{3.3}$$

Hence the curve shown in Fig. 3.8 for the flexural mode is part of a rectangular hyperbola.

Figure 3.8 shows the two curves for the thickness and flexural modes intersecting at a specific value of L/h. As already indicated, the basic assumption of the previous discussion is invalid at this point. A more exact analysis, supported by experiment, shows that if the relevant mode frequencies of a strip resonator are plotted as a function of L/h, with L being gradually reduced by, for example, grinding the edges of the resonator, the actual behaviour is as shown in Fig. 3.9. Here at the point A on the thickness mode curve, the flexural mode is too far away in frequency to have any effect. The particle displacements are predominantly thickness shear, as in Fig. 3.2. Moving down the thickness mode curve to point B, the flexural mode begins to influence the character of the displacements, and the frequency is perturbed upwards. Moving still further along the same curve, the character of the displacements changes to predominantly flexural, until finally a complete transition is made from thickness mode to flexural mode behaviour at point C. In the same way, starting at the point A' on the flexural mode curve and proceeding through the points B' and C', the mode character changes from flexural at A', through a combination of flexure and thickness shear at B', to purely thickness shear at C'. Both curves thus show a

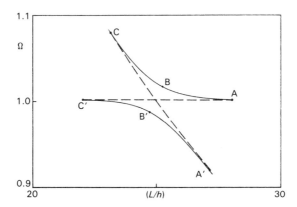

Fig. 3.9 Effect of coupling on TS and F modes.

continuous change in character, and do not intersect at the critical L/h value as predicted by the simple arguments used previously. The two modes in question are said to be *coupled*. The question of the exact mechanism of the coupling is begged by the discussion so far, but it suffices for present purposes to recall the statement made in Section 3.2 that satisfaction of the edge conditions requires a superposition of plate waves; if the dimensions of the resonator are such that only one mode satisfies the resonance conditions, then it is physically reasonable to suppose that this mode will predominate in the solution. However, if two or more modes simultaneously satisfy the resonance conditions, then clearly both modes must be expected to have significant amplitudes. Thus it can generally be supposed that where two modes can be simultaneously excited into resonance, they will be coupled together through the boundary conditions. The obvious exception to this rule would be the case where one or both modes could satisfy the edge conditions independently of the other, but as already indicated this is a rare situation.

The previous paragraphs apply equally well to both piezoelectric and non-piezoelectric resonators. In the former case, some further conclusions can be drawn based on the general character of the displacements in the thickness and flexural modes. Firstly, in the thickness mode case, when no coupled modes are present, the displacements are such that to a first approximation all points on each major face are moving in phase. Consequently, supposing the vibrations to be excited by electrodes on the major surfaces, the piezoelectric charges produced on the electrodes will likewise be in phase. On the other hand, in the case of the flexural modes at frequencies near the thickness mode, the wavenumber will be large (approximately L/h times the thickness mode wavenumber), and so there will be a large number of phase reversals from one edge of the plate to the other. The piezoelectric charges will therefore tend to cancel out, resulting in a relatively weak excitation of the flexural mode. In equivalent circuit terms, this implies that the effective resistance of the flexural mode is much higher than that of the thickness mode. Now if the resonator is used in a practical oscillator circuit, the preferred mode of vibration will be the one with the lower effective resistance. Hence if the output frequency of the oscillator is plotted as a function of the L/h ratio, the typical behaviour would be as shown in Fig. 3.10. Here at point A, corresponding to point A of Fig. 3.9, the resonator is working in its relatively low resistance thickness mode. As L/h is reduced and the coupling to the flexural mode increases, the oscillator frequency is 'pulled' upwards and at the same time the effective resistance increases. As L/h is further decreased, the effective resistance continues to increase, until at point B say, the condition is reached where the resistance at B' is less than at B. Then the oscillator 'jumps' in frequency from B to B', and with further reduction in L/h, continues along the curve through the point C'.

Although specific mention has been made only of coupling between the thickness shear and flexural modes, similar coupling exists between the

ENERGY TRAPPING

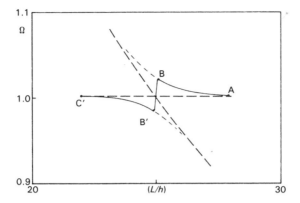

Fig. 3.10 Effect of coupling on oscillator frequency.

desired mode and other modes such as face shear and extensional. In all cases, the effects of such coupling are detrimental. At the end of the previous section it was pointed out that the temperature characteristics of the flexural and face shear modes are quite different to those of the thickness modes. Consequently, for resonators operating over wide temperature range, it can easily happen that although a particular flexural or other unwanted mode may be far removed from the desired thickness mode at one temperature, it may nevertheless coincide with the wanted mode at some other temperature in the range. When this occurs, coupling of the modes will produce frequency jumps at the particular temperature in question, usually accompanied by sharp increases in the crystal resistance. These effects of temperature-dependent coupling are variously known as 'activity dips' or 'bandbreaks'.

In the early days of the manufacture of AT- and BT-cuts, the need to minimize such coupling resulted in the introduction of elaborate dimensioning rules governing the allowed ranges of the length-to-thickness and width-to-thickness ratios. These resulted in plates of only slightly different frequencies having to be ground to different lengths and widths, considerably complicating the manufacturing process as compared to present-day techniques.

3.4 ENERGY TRAPPING

The key concept of the cut-off frequency for thickness twist and thickness shear waves in plates has already been introduced (Section 3.1 and Appendix 6). Consider for definiteness thickness twist waves in an isotropic plate, with thickness $2h$ along the y direction, particle displacement u along the x direction, and propagating along the z direction. From Appendix 6, the Nth odd overtone mode has the displacement

$$u = A\sin(N\pi y/2h)\exp(j\omega t - jk_z z) \tag{3.4}$$

For real k_z this represents a travelling wave, whereas for an imaginary k_z, say with $k_z = j\bar{k}_z$, the expression becomes

$$u = A\sin(N\pi y/2h)\exp(\bar{k}_z z)\exp(j\omega t) \tag{3.5}$$

representing an evanescent wave.

Now consider a plate as above except that in the region $|z| < W$ the thickness is $2H$, with $H > h$ (Fig. 3.11). Then in the central region the cut-off frequencies are lowered in the ratio h/H. Thus if the dispersion curves for both the central and outer regions of the plate are plotted on the same axes, the result will be as in Fig. 3.12, where only the Nth TT mode is shown for

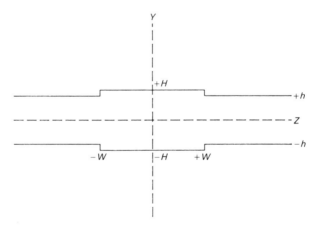

Fig. 3.11 Geometry of energy trapped resonator.

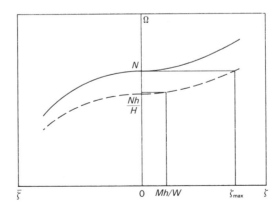

Fig. 3.12 Dispersion relations for energy trapping.

clarity. For normalized frequencies $\Omega > N$, that is above the cut-off for both regions, TT waves can propagate in both central and outer areas. Similarly, for $\Omega < Nh/H$, only evanescent waves can exist. However, in the frequency range $Nh/H < \Omega < N$, propagating waves can exist in the central region but not in the outer regions. This immediately gives rise to the possibility of producing a resonator by utilizing the reflections of the waves at the edges of the central region to set up a standing wave system. Such a resonator is termed a *trapped energy resonator*, since the vibrational energy is 'trapped' in the central region of the device.

The width $2W$ is the critical factor in determining the resonance frequencies of the trapped modes, via the general assumption that resonance will occur when an integral number of half wavelengths are contained in the critical dimension. In terms of the normalized wavenumber $\zeta = 2hk_z/\pi$, this means that the relevant ζ values are

$$\zeta = Mh/W \tag{3.6}$$

where M is an integer. Eqn (3.6) is exactly analogous to Eqn (3.1) in Section 3.2, except for the fact that although the integer N in Eqn (3.1) can take on all values, the corresponding integer M in Eqn (3.6) can only take on a finite number of values. The reason for this can be seen from Fig. 3.12, where the points Mh/W are marked off on the ζ axis. Clearly for all M greater than some value M_0, the mode frequency will be above the upper limit N, so that those modes will no longer be trapped.

The number of trapped modes depends both on the width $2W$ and the difference between the cut-off frequencies in the two regions. If Δ is the fractional difference between these frequencies

$$\Delta = (H-h)/h \tag{3.7}$$

Also the dispersion relation in the central region is

$$\Omega^2 = (Nh/H)^2 + \zeta^2 \tag{3.8}$$

The maximum value of ζ for a trapped wave is obtained when $\Omega = N$, the cut-off frequency for the outer region, so that

$$\zeta_{max}^2 = N^2[1 - (h/H)^2]$$

or, in the usual case where $(H-h)/h \ll 1$,

$$\zeta_{max}^2 = 2N^2\Delta \tag{3.9}$$

The condition for there to be just M trapped modes is that the maximum wavenumber $\zeta_{max} = Mh/W$, that is

$$(Mh/W)^2 = 2N^2\Delta \tag{3.10}$$

Writing B_M for the ratio W/h, Eqn (3.10) can be rewritten as

$$B_M = M/(2N^2\Delta)^{1/2} \tag{3.11}$$

and for a given Δ, determines the width-to-thickness ratio required to selectively trap the first M thickness twist modes at the Nth overtone.

3.5 MASS LOADING

Although the preceding discussion has been based on the specific case of a plate with a thick central region and a thinner outer region, the phenomenon of energy trapping only requires that there be a difference in cut-off frequencies between the two regions. The precise mechanism by which this difference is achieved is immaterial. In practice, the example considered of a step change in the plate thickness is not used, although such a change could be achieved by selective etching of the crystal blank. The two commonly used techniques for achieving different cut-off frequencies in different regions of the resonator employ in the one case the frequency lowering due to the mass of the electrodes, and in the other a gradual change in thickness obtained by *contouring* or *bevelling* the crystal blank.

The *mass loading* effect results from the modification of the boundary conditions on the major surfaces of the resonator due to the presence of the electrodes. For major surfaces at $y = \pm h$, the stress free boundary conditions used in Chapter 2 and Appendix 6 were simply the vanishing of the surface tractions t_{2k}. In the presence of electrodes of mass density σ per unit area, the inertia of the electrodes results in the boundary conditions.

$$t_{2k} = -\sigma \ddot{u}_k \qquad (3.12)$$

The relative magnitude of the mass loading is expressed in terms of the ratio R of the electrode mass per unit area to the mass per unit area of the resonator, $R = \sigma/\rho h$. For large R, the use of Eqn (3.12) rather than $t_{2k} = 0$ in the analysis of thickness modes in Chapter 2 leads to frequency equations best solved graphically or numerically (Ballato, 1977). However, it is usually the case with quartz resonators that the mass loading is small, with $R \ll 1$, and then the modified boundary conditions simply lead to the result that the resonance frequencies for thickness modes are lowered by the fractional amount R. That is, if f_m and f are the frequencies with and without mass loading

$$(f - f_m)/f = R \qquad (3.13)$$

Again in the case of small R, this result can be carried over to the dispersion relations for wave propagation in mass loaded plates, insofar as the thickness shear and thickness twist branches, at least in the neighbourhood of zero wavenumber, can be regarded as having the same shape as for unloaded plates but shifted down on the frequency axis by the fractional amount R.

Consequently, if a resonator in the form of a thin plate or disc with a central region coated with electrodes is considered, then the mass loading of the electrodes will result in a finite number of TT and TS modes being trapped under the electrodes. The number of such modes will be given by an analogue of Eqn (3.11) with Δ replaced by R. The precise values of the numerical constants involved will of course differ from those in Eqn (3.11), since the analysis leading up to this equation only considered the simplest case of TT waves in an isotropic plate. Nevertheless, the essential points remain that the number of trapped modes increases both with the frequency lowering, in this case the mass loading R, and the size of the electrode. In particular, the possibility exists to limit the number of trapped modes to one only by judicious control of the mass loading and the electrode dimensions. This was first recognized in practice by Bechmann and the critical ratio of electrode dimension to plate thickness required to trap only one mode is known as *Bechmann's number* (for references, see Meeker, 1985).

The practical importance of this is that the energy of the trapped mode, being concentrated under the electrodes, gives rise to a strong resonator response, whereas the modes that are not trapped can propagate over the whole of the resonator. Thus not only is the proportion of the total energy in the mode that is available to the driving circuit restricted, but also these untrapped modes are subject to the usually large damping that occurs at the edges of the resonator plate, for example at the mounting points. The net effect of these two factors is that a properly designed resonator using the energy trapping criteria will have a much 'cleaner' response than other resonators, that is, it will be essentially free from 'unwanted' or 'spurious' responses.

3.6 CONTOURING AND BEVELLING

The second commonly used technique for using the energy trapping phenomenon, that of contouring the crystal to give a gradually changing thickness from the centre to the edge of the blank, was a well established empirical practice (Tyler, 1960) long before its theoretical analysis. The effect of contouring, that is the practice of imparting a spherical curvature to one or both major faces of a resonator as shown in Fig. 3.13, was recognized to be the restriction of the vibrating area of the plate to its central region, with the accompanying advantages of ease of mounting and reduced coupling to unwanted modes at the edge of the blank. Also recognized, but not understood on a theoretical basis, was the pronounced change in the first-order frequency–temperature characteristic brought about by contouring. Empirical data, in the form of curves and tables of 'good' designs, was, as stated above, well established for contoured units by the early 1960s. Similar data for *partially contoured* or *bevelled* blanks, where the spherical curvature is only applied in the outer region leaving a flat central region or *plateau* as in

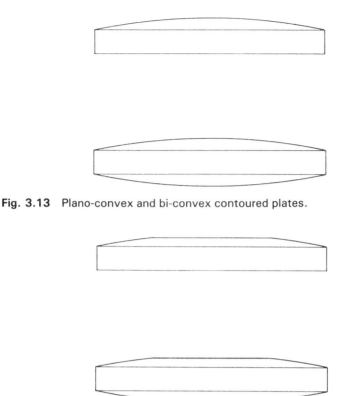

Fig. 3.13 Plano-convex and bi-convex contoured plates.

Fig. 3.14 Plano-convex and bi-convex bevelled plates.

Fig. 3.14, was not, however, readily available, and information on such designs is even now only available on a restricted basis.

The analysis of Section 3.4 is not directly applicable to contoured resonators, but nevertheless helps to give a physical understanding. A precise analysis (Tiersten and Smythe, 1979) shows that for the main response of a spherically contoured resonator, the amplitude of vibration falls away exponentially with the *square* of the distance from the centre, rather than as a simple exponential function of distance. This is to be expected on physical grounds since the cut-off frequency itself is a function of distance from the centre. Tiersten and Smythe give explicit solutions for both the main response and the inharmonic responses of a fully contoured resonator, from which the frequencies and the equivalent circuit parameters of each mode can be calculated. To the author's knowledge, no equivalent exact solution is available for the partially contoured or bevelled blank, except in the relatively simple and rarely used case of a *cylindrical* as opposed to a spherical contour. Thus to obtain solutions for the bevelled case, recourse has to be had to numerical methods.

Part 2
Manufacturing techniques

4 Optical processing

The manufacture of any type of quartz resonator can be divided into two phases. In the first phase, traditionally known as *optical* processing because of close similarities with the manufacture of optical components such as lenses, the operations are directed towards the production of crystal blanks of the desired orientation, dimensions, shape and surface finish. The second phase, known variously as *electrical* processing, or simply *finishing*, covers the operations of cleaning, laying down electrodes (usually by vacuum deposition), mounting, adjusting to frequency and sealing. These process steps are described in more detail in the following chapter.

Figure 4.1 shows a flow chart of the main steps involved in the optical processing of thickness mode resonators such as the AT-cut. The steps are:

(1) Orientation of the crystal bar before cutting.
(2) Cutting or sawing the bar into *wafers*.
(3) Rough lapping of the wafers and angle sorting.
(4) Removal of the seed.
(5) Rounding circular blanks (edging rectangular blanks).
(6) Fine lapping
(7) Angle sorting

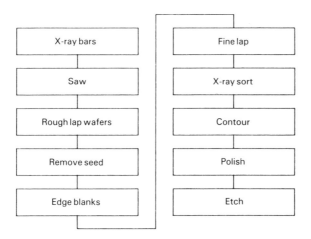

Fig. 4.1 Optical processing steps.

(8) Polishing.
(9) Contouring or bevelling.
(10) Etching.

Depending on the frequency range and performance requirements of the finished resonators, some of these operations may be omitted. The etching process is often physically located in the finishing area, or alternatively in a separate area between optical and finishing shops. Consequently, it is discussed in the following chapter as part of the blank cleaning operation, although strictly speaking it should be included as an optical process since it directly affects both dimensions and surface finish of the crystal blank.

4.1 X-RAY ORIENTATION

For the commonly used AT-cut crystal, tolerances on the angle at which the blank must be cut from the parent crystal are typically of the order of minutes of arc. Increasingly in modern communication systems the frequency stability requirements are such that even tighter tolerances of less than a minute are required. Although in the early days of the industry optical means of orientation were often used, *X-ray diffraction* is now universally adopted as the appropriate technique for achieving the required accuracy.

The physical principles of X-ray diffraction by crystals are the same as those of optical diffraction, with the proviso that in the optical case the wavelengths involved are much longer. In both cases, a beam of radiation is incident upon an array of scattering centres which is periodically arranged in space, and constructive interference between the scattered radiation from the individual elements of the array results in relatively intense scattered or *diffracted* beams in certain directions relative to the array and to the direction of the incident beam. In the optical case, the scattering centres may be the lines of a diffraction grating, whereas in the present case the centres are the individual atoms and molecules of the crystal structure. In either case, the wavelength of the radiation has to be less than or of the order of the spacing of the array elements. For quartz, the unit cell dimensions are approximately 5×10^{-10} m, hence wavelengths of this order are necessary to observe any diffraction phenomena.

If only the angles of the diffracted beams are of interest, rather than their relative intensities, the simple concept of the reflection of X-rays from *atomic planes* in the crystal can be substituted for the more general notion of diffraction. In any crystal, because of its periodic structure, any plane drawn through specific atomic or molecular locations in the crystal will have associated with it an infinite number of similar parallel planes. This family of planes will be characterized by a definite spacing d between successive members. Each plane will intersect the crystallographic axes at certain points, except in the case where the plane is parallel to a particular axis or axes, when the intercepts would be at infinity. If one particular plane that intersects all three axes is chosen as a reference plane, and the intercepts of this *parametral* plane on the axes are taken as units in which to express the intercepts of any other plane, then it is a fundamental law of crystallography (Phillips, 1960) that the reciprocals of the intercepts when so expressed are rational numbers or zero. This is the *law of rational indices*, the reciprocals of the intercepts being the *indices* of the plane in question. Supposing that a given plane has indices (h, k, l), it is clear that the indices (Nh, Nk, Nl) represent a parallel plane, and thus if only the orientation of a family of planes is of interest, it can be represented by a set of three integers obtained by multiplying the indices (h, k, l) of any member of the family by some integer N to clear fractions. When this is done the indices are termed the *Miller* indices of the family of planes, and conventionally written (hkl) without intervening commas. In the case of quartz and other trigonal or hexagonal crystals, it is conventional to use a coordinate system with the Z axis along the trigonal or hexagonal axis, and *three* equivalent X axes labelled X_1, X_2 and X_3 perpendicular to the Z axis rather than the normal X and Y axes. Then the system of Miller indices is modified to the *Bravais–Miller* system using four indices $(hkil)$, where the indices refer respectively to the reciprocals of the intercepts on the X_1, X_2, X_3 and Z axes. However, because all three X axes are coplanar, the first three indices satisfy the identity $h + k + i = 0$, and the Bravais–Miller indices are often written $(hk.l)$ to emphasize this dependence.

The essential ideas of X-ray reflection are first that X-rays are reflected from atomic planes just as light beams are reflected from plane mirrors, that is, with equal angles of incidence and reflection, and second, that *strong* reflections only occur at those angles of incidence such that the reflections from all the planes parallel to the given plane reinforce each other. This condition is expressed in Bragg's law of X-ray reflection

$$2d\sin(\theta) = n\lambda \qquad (4.1)$$

where θ is the glancing angle or *Bragg angle*, λ is the wavelength of the X-ray beam, d is the spacing of the atomic planes, and n is an integer. Bragg's law simply expresses the condition that the path difference between reflections from successive planes be an integral number of wavelengths; Fig. 4.2 illustrates the geometry involved.

Equation (4.1) provides the basis for the use of X-rays in determining crystal orientations. The first step is to select an atomic plane which lies parallel or almost parallel to the desired orientation of the major surfaces of the crystal. In the case of the AT-cut the plane normally selected has Bravais–Miller indices (10.1), for which the interplanar spacing is 3.3362 Å. Assuming that the wavelength λ is 1.537395 Å corresponding to the Cu Kα_1 line, the first order reflection, $n = 1$, occurs at the Bragg angle θ_B of 13° 19'. If now a collimated X-ray source and detector are arranged as in Fig. 4.3,

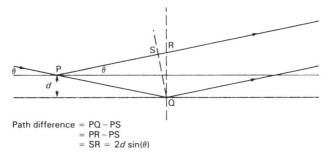

Fig. 4.2 Geometry of X-ray reflection.

such that the angle between them is $2\theta_B$, then it is clear that a strong reflection will only be obtained when the crystal specimen at O is arranged such that the selected plane is in the position shown.

The principle of an X-ray goniometer is then to set up source and detector in fixed positions as above, and provide a means for rotating the crystal specimen about the vertical axis through O in Fig. 4.3 until a strong reflection is obtained. By measuring the angle between the surface of the specimen from which the X-rays are being reflected and the reference line AA', the angle between the surface and the selected atomic plane can be determined. The procedure is in detail dependent on the choice of the atomic plane, and is particularly simple in the case of AT cuts if the (10.1) plane is used because the intersection of this plane and the crystal surface is parallel to the vertical axis of the goniometer. In other cases, several measurements involving different atomic planes may have to be made before the orientation of the plate can be fully determined. These procedures are described in Heising (1946), Bond (1976), and Bottom (1982). Figure 4.4 shows a typical goniometer system incorporating a double reflection arrangement to provide greater precision than the single reflection arrangement outlined above.

In the manufacturing process, X-ray procedures are used both to set up the crystal bars for cutting and subsequently to check the orientation of finished or semi-finished blanks. Cultured quartz bars which are now almost univer-

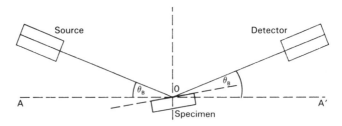

Fig. 4.3 X-ray goniometer arrangement.

Fig. 4.4 X-ray goniometer.

sally used are supplied by the manufacturers with carefully prepared reference surfaces ground normal to the crystallographic X axis and typically have their length and width along the Y and Z axes, respectively, as in Fig. 4.5. The major surfaces of *lumbered* bars are ground flat and mutually perpendicular, but the ends of the bar show the 'natural' crystal faces, and either one of these natural faces or a lumbered Z surface can be used to set up the bar for cutting. For this purpose the bar is usually mounted on a 'transfer' jig which can be removed bodily from the X-ray machine and loaded directly onto the saw for cutting.

For checking the orientation of finished or semi-finished blanks, vacuum chucks are used to hold the blanks in position against a reference surface on the chuck. The reference may be provided either by a precision ground and lapped surface, or for greater accuracy and repeatability, by a three-point mounting system consisting of ruby tubes carefully set into the chuck surface so as to provide a reference plane perpendicular to the plane of the instrument. A high degree of cleanliness is essential in this operation, since even small dust particles between the blank and the reference surface of the chuck can cause unacceptable errors. For example, a 1 micron particle under one edge of a 5 mm diameter crystal blank will cause an angle error of approximately 0.7 minutes of arc.

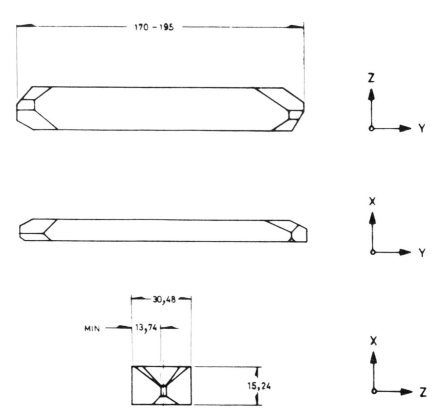

Fig. 4.5 Lumbered quartz bar.

It should also be noted that precision X-ray measurements are wasted unless the surfaces of the blanks to be measured are flat and parallel to a sufficient degree. If tolerances of less than a minute of arc are to be achieved, it is obvious that departures from parallelism of more than a small fraction of a minute cannot be tolerated. Of course, for contoured blanks, the major surfaces are by design not parallel. For plano-convex units, the plano side can still be used for X-ray measurements, and then the critical factor becomes the location of the centre of curvature of the contoured side relative to the plano side. For bi-convex blanks, no reference surface is available for X-ray measurements on the finished unit, and the critical factors are the relative locations of the centres of curvature both to each other and to the original central axis of the initially flat plate. Because of these considerations, it is in practice impossible with biconvex crystals to maintain such close control over orientation as is routine with flat blanks.

4.2 SAWING, DICING AND ROUNDING

Several different methods are in use for sawing quartz bars into wafers. One of the most popular is the 'slurry saw' technique, which is in fact more a lapping process than a sawing one. Figure 4.6 shows a typical saw. One or more quartz bars are rigidly mounted on the bed of the machine, precisely oriented by means of X-ray transfer jigs as previously mentioned. The saw blades consist of tensioned metal strips made up into *blade packs* containing up to a hundred or more blades depending on the desired wafer thickness. The blade pack is then driven in a reciprocating motion along the length of the fixed bed of the machine, back and forth over the quartz bars, with the whole continually being flooded by an oil-based abrasive slurry. The sawing operation takes several hours, but since the machines can be left to run unattended, and are capable of cutting several bars simultaneously, the productivity of the process is high.

The wafers resulting from the sawing process are rectangular in shape, and each contains in its central region a portion of the original seed crystal used in the growth of the bar, which cannot be used owing to the high density of defects and impurities found in the immediate neighbourhood of the seed. Thus the central region of the wafers has to be removed. The wafers are

Fig. 4.6 Slurry saw.

therefore first lapped to remove any wedging that may have resulted from the sawing operation, and then waxed or cemented into long stacks. The stacks are then sawn into two stacks of roughly square *dice*, the central region being discarded. Depending on the type of resonator required, the next operation is then either to surface grind the dice, still in their stacks, to specified edge dimensions, or else to *circularize* or *round* them to a specified diameter. The rounding operation may be carried out on conventional cylindrical grinders or on centreless grinders, the latter generally requiring less fixturing and giving a better edge finish. However, the centreless grinder can only produce a completely circular stack, and it is often desirable to retain a small flat perpendicular to the X axis on the blanks to aid orientation at subsequent stages. With appropriate fixtures this is straightforward to achieve on a conventional grinder, by offsetting the axis of the stack relative to the axis of the machine, but requires an additional operation when using a centreless grinder.

After the edging or rounding operations, the dice or blanks are separated and cleaned ready for subsequent operations. At this stage, the surface finish is relatively coarse, and the blanks are relatively thick, it being impractical to cut wafers much thinner than two to three hundred microns.

4.3 LAPPING AND POLISHING

For low frequency resonators such as extensional, flexural and face shear types, the blank frequency is essentially determined by the lateral dimensions of the blank, with the thickness governing the motional inductance. For thickness mode resonators, the reverse is true, the thickness being the main determinant of the blank frequency. For the AT-cut, in the fundamental frequency range 1 to 50 MHz, the thickness ranges from about 2 mm down to 33 microns, with the tolerances required being of the order of 0.1%. There are accompanying severe requirements on blank flatness and parallelism. The surface finish required can be assessed in terms of the acoustic wavelength in the resonator, as it seems physically reasonable to assume that to avoid losses due to surface irregularities, such surface damage must be restricted to a small fraction of a wavelength. This immediately leads to the conclusion that the degree of surface finish required increases with frequency, and in particular that overtone resonators require a higher degree of finish than a fundamental mode resonator of the same thickness.

In order to achieve the required thickness and surface finish, a graded sequence of double-sided lapping operations is commonly used, in some cases ending with a polishing operation in which the blanks are given an optical polish. When large amounts of material have to be removed, one of the double-sided lapping stages may be replaced by single-sided lapping,

LAPPING AND POLISHING 71

Fig. 4.7 Planetary lap.

which has the advantage of better control over the orientation angle but the disadvantage of not maintaining parallelism. Hence double-sided lapping is always used after any single-sided operation.

Double-sided lapping machines are of two main types, *planetary* and *eccentric* or *pin* laps. In both types, the blanks are carried around between cast iron lapping plates in an oil or water based abrasive slurry. In the planetary system (Fig. 4.7) the lapping carriers have the form shown in Fig. 4.8, and are located between toothed inner and outer gears. The gear

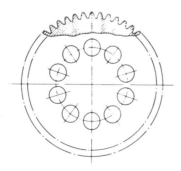

Fig. 4.8 Planetary lap carrier.

ratios are so chosen that in the motion of the blanks, the surfaces of the plates wear uniformly. This helps to maintain both the flatness of the plates themselves, and also ensures a corresponding flatness in the blanks, but nevertheless the lapping plates have frequently to be removed and lapped flat on a larger machine to guarantee blank quality. Blank parallelism is ensured by regular *transposition* of the blanks in the lapload. Transposition is a systematic interchange of blanks from opposite sides of the lapload designed to prevent a situation from arising where all the thicker blanks are grouped together on one side of the load. If such a situation does occur, in which the upper lap plate is resting on the blanks at some small angle to the lower plate, then there is no intrinsic correcting mechanism and without transposition to redistribute the thicker blanks, the result would be both an increasing spread in frequency between the blanks in the load and a pronounced 'wedging' of the blanks.

In the pin or eccentric lap (Fig. 4.9) the lapping principles are the same, but

Fig. 4.9 Pin lap.

LAPPING AND POLISHING 73

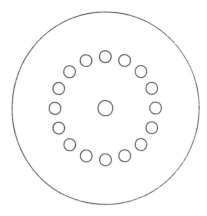

Fig. 4.10 Pin lap carrier.

the mechanism by which the blanks are driven around between the lapping plates differs. The blanks are held in a carrier (Fig. 4.10) consisting of a sheet of 'mylar' film or similar material, supported in its centre, and free to rotate about a pin that is itself located slightly off centre relative to the vertical axis of the machine (Fig. 4.11) In operation, this eccentricity of the carrier centre once again results in a motion in which each individual blank eventually traverses all points on the surfaces of the lapping plates, with the object being to preserve the flatness of both lap plates and blanks. This ideal is not of course achieved in practice, so that as in the planetary case, the lap plates have frequently to be removed for reconditioning.

In both types of lap it is common to use some means of automatic control to terminate the operation when the desired thickness or frequency has been reached. The traditional technique is simply to detect the transient

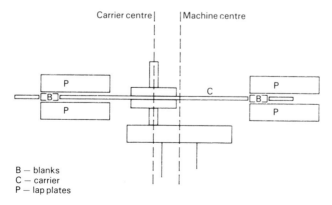

B — blanks
C — carrier
P — lap plates

Fig. 4.11 Cross-section of pin lap geometry.

oscillations of the blanks excited by the lapping process using a sensitive radio receiver connected to one of the plates as an aerial. These oscillations appear as a noise signal centred on the blank frequency, and can either be used to monitor the progress of the operation or to provide an automatic shut-off by setting the receiver to a target frequency. A more sophisticated technique is to insert electrodes into both upper and lower lapping plates and to monitor the change in impedance that occurs when a blank passes between the electrodes. A swept frequency RF signal is applied to the electrodes and during each sweep the frequencies at which impedance changes occur are stored. At the end of the sweep, the mean or maximum frequency stored, together with the spread in the stored frequencies, can be displayed, and the machine shut off when a preset target frequency is reached.

The abrasives used in the lapping operations are generally carefully graded aluminium oxide powders, with silicon carbide often being used for the coarser stages, and sub-micron cerium oxide powders being used for polishing. The coarser the abrasive, the more rapid the stock removal but the poorer the surface finish, so it is usual to grade the lapping operations. In the initial stages a coarse abrasive is used, and in subsequent stages, successively finer materials. In each stage, the amount of material removed is related by rule of thumb to the size of the previous abrasive, so that at the end of each stage the surface finish of the blanks is characteristic of the abrasive used in that stage rather than a previous one. Polishing operations differ little in principle from lapping operations. The abrasive used is much finer, and in addition the lapping plates are usually covered with a pad of some material. The particular material used varies from manufacturer to manufacturer, but as a general rule, soft pads quickly produce a high polish with relatively coarse abrasives but do not maintain flatness and parallelism, whereas harder pads have the reverse characteristics.

4.4 CONTOURING AND BEVELLING

Low frequency thickness mode resonators such as the AT are frequently given a complete or partial spherical contour on one or both major surfaces, in order to restrict the vibrating area of the blank to its central region and thus avoid mounting losses and coupling to unwanted modes at the edges of the blank. Contouring methods fall into two distinct classes, one suited to relatively high volume production, the other to small batch production.

The first group comprises those techniques commonly referred to as *tube* or *drum* contouring, in which the mechanism involved is very similar to that responsible for the rounding of pebbles in a river bed. The procedure is to load a quantity of blanks, previously lapped to the desired final thickness, together with a quantity of abrasive powder, into a pipe or drum of internal

Fig. 4.12 Contouring machine.

radius equal to the desired radius of curvature for the blanks. The tube and its contents are then continuously rotated about the tube's longitudinal axis in such a way that the blanks are subject to a gentle tumbling action, which over a period of perhaps several days results in the blanks taking on the contour of the tube. Naturally both surfaces are symmetrically affected, so that the contour is a biconvex one, and depending upon the time allowed in the tube, will either be complete or a partial contour (or bevel). The process is essentially statistical in nature, that is not all blanks will attain the same degree of contour in the same period of time, so that periodic monitoring is essential.

Although planoconvex contours can be produced in tube contouring by the expedient of cementing blanks together in pairs, such contours are usually produced by methods derived from the optical industry, which comprise the second of the two classes mentioned above. Figure 4.12 shows a typical contouring machine using a 'dioptre cup'. This is a cast iron tool with a concave spherical surface machined on its upper face, and designed to be mounted on a vertical spindle whose axis goes through the centre of curvature of the concave surface. In operation the tool is rotated about this vertical axis and the crystal blank, usually mounted on a brass 'contouring button', is swept backwards and forwards across the cup, from a point near its edge to a point just past its centre, with the whole being flooded by an abrasive slurry.

The blank can either be held manually or, as in the illustration, held by a mechanical arm locating in a recess machined in the back of the button. In the manual case, the blank must be continually rotated about its own axis to maintain the spherical character of the contour; in the mechanical case, the rotation required is generated by the rotation of the cup, provided only that the button is free to spin about the axis of the mechanical arm. Clearly this process is intrinsically better suited to small batch production rather than high volume work, but nevertheless is extensively used in all classes of production. Consequently the cost savings that result from volume production in the higher frequency ranges cannot necessarily be achieved at low frequencies.

5 Electrical processing

5.1 CLEANING AND ETCHING

During the optical processing stages, crystal blanks are subject to a wide variety of soils and contaminants, all of which must be removed before the following process stages. In addition, by the nature of the lapping process, however fine an abrasive is used the surface of a lapped blank would still show on examination under sufficiently high magnification a fine structure of minute irregularities. These would include myriads of fine cracks and fissures, accompanied by occasional deeper scratches from abrasive particles with larger than average diameter, from dust particles, or from other foreign matter. Even in the case of (mechanically) polished blanks, it seems probable that a damaged layer from previous lapping stages still exists under the polished surface.

This lack of surface perfection has the immediate consequence of degrading the Q of the resonator, with the depth of the surface damage compared to the acoustic wavelength at the operating frequency giving some idea of the degree of degradation. Over a period of time, a less immediately obvious consequence of surface imperfections shows itself in a rapid upward drift of the resonator frequency caused by the physical loss of material from the resonator surfaces. This is the result of small particles of quartz having been loosened in the lapping process and subsequently separating completely from the main body of the material. To prevent this ageing it is necessary to remove any such loose particles by *etching* away the surface layers of the blank to obtain a more stable surface. This etching process at the same time improves the resonator Q by reducing acoustic losses at the surfaces.

The amount of material removed in the etching process is usually related by rule of thumb to the size of the abrasive used in the final lapping stage. However, in recent developments the application of the etching technique has been extended so that the process of *deep etching* or *chemical polishing* has become an alternative to the traditional mechanical polishing procedures. In chemical polishing the amount of material removed is one to two orders of magnitude more than in conventional etching, and chemically polished blanks are claimed to have Q factors as good as or better than mechanically polished units, and also to be mechanically more robust. Deep etching does however make more stringent demands on raw material quality, since the presence of certain defects can lead to deep etch pits and in some extreme cases to etch channels extending throughout the thickness of the blank.

Various etchants have been proposed but the most regularly used have been aqueous solutions of ammonium bifluoride or hydrofluoric acid. Most often these solutions are used hot to speed up the process, and in all cases the blanks need to be thoroughly rinsed to remove all traces of etchant before being passed on to the next processing stage. For the etching to proceed in a uniform manner the blanks must be thoroughly cleaned beforehand, and it is usual to integrate the cleaning and etching procedures into one sequence of operations.

Different manufacturers each have their own preferences, but a typical clean and etch sequence would include the following steps:

(1) Load blanks into cleaning jigs fabricated from stainless steel or other inert material. The jigs should be so designed that the maximum blank area is exposed to the cleaning process.
(2) Ultrasonic wash in any one of several alternative solutions or solvents, including detergents, alcohols, acetone, trichloroethylene, and acidic or alkaline baths. Sometimes a sequence of washes may be used.
(3) Thoroughly rinse blanks in deionized water.
(4) Etch in saturated ammonium bifluoride solution.
(5) Thoroughly rinse off all traces of etchant.
(6) Dry blanks, using any one of several alternative techniques such as spin drying, hot air blowers, or chemical displacement methods. The latter include simple procedures involving successive alcohol and acetone rinses followed by blow drying, and more sophisticated techniques using proprietary equipment based on blends of trichlorotrifluorethane.

However elaborate the washing sequence used, it seems impossible to remove all traces of contaminants by conventional methods. In recent years new techniques have been introduced to complement the traditional procedures, with the most effective appearing to be the exposure of the blanks to intense ultraviolet radiation. In the presence of oxygen, the UV radiation generates ozone, which is critical to the effectiveness of the cleaning method, hence the usual term *UV-ozone* cleaning (Vig, 1971, 1975). When used, this step is carried out after a conventional wash sequence and immediately before the deposition of electrodes.

5.2 BASE-PLATING

All resonators require some form of electrode structure. In early devices *air-gap* electrodes were used, consisting of metal plates separated from the surfaces of the resonator by a small gap. Later, electrodes deposited directly on the quartz surfaces by vacuum deposition became the universally accepted standard, and the overwhelming majority of currently produced crystal units still use this technique. Nevertheless, for resonators of the very highest precision and stability, air-gap electrodes are once more being used to avoid the instabilities associated with electrode stresses, so that the wheel has come full circle for these devices (Besson, 1976).

In the conventional approach using vacuum deposition, the shape of the electrodes is usually defined by using photo-etched plating masks. These are commonly used in a 'sandwich' arrangement of three layers held in a rigid frame to allow ease of handling. The sandwich consists of a mask to define the electrodes on one face, a spacer with nests in which to locate the crystal blanks, and a second mask to define the electrodes on the opposite face. The material used to fabricate the masks is generally stainless steel. After loading the crystal blanks into the masks, with each mask generally holding a large number of blanks, several masks are loaded into carriers in a vacuum chamber. Depending upon the design of the unit, provision is usually made for the carriers to rotate around the filaments during the plating operation so as to ensure uniform plating from blank to blank and mask to mask. Also it is usual to arrange for the masks to be automatically turned over on completion of plating on the first side of the blanks, so that the second side may be plated without breaking the vacuum in the system. Figure 5.1 shows the interior of a typical base-plating machine.

A high vacuum is necessary in the chamber to ensure the quality of the deposited film. This can be attained by using various different pumping systems, including oil diffusion pumps, turbomolecular pumps, ionization pumps, and cryogenic pumps. Diffusion pumps have been widely used in the past, but the modern trend is towards the use of cryogenic pumps, largely because of their ease of use and their inherent cleanliness.

The most commonly used electrode materials are silver, gold and aluminium, with copper having been used in some special applications. It is common to use a thin layer of chrome under silver or gold electrodes in order to improve the adhesion of the electrodes to the quartz, but this has the disadvantage of increasing the stresses in the electrodes and thereby providing additional sources of long-term instability. The electrode materials can be deposited either by sputtering or by evaporation; the latter being most popular. The thickness of material deposited is controlled by detecting the change in frequency of a monitor crystal exposed to the evaporant, the whole plating cycle usually being fully automated.

5.3 MOUNTING AND BONDING

The basic requirements in mounting crystal resonators are that the crystal be provided with mechanical support, with protection from adverse environ-

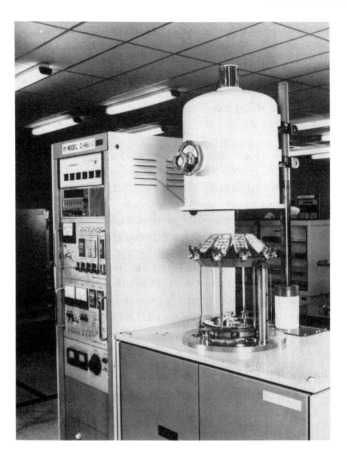

Fig. 5.1 Base plater.

mental effects (humidity, corrosive atmospheres, airborne contaminants, etc), and with some means of electrical connection to the driving circuit. At the same time, the mounting structure should impose minimal stresses on the resonator and minimize the damping of the mechanical resonance (Ward, 1983). The vast majority of medium to high precision resonators produced are packaged in hermetically sealed metal, glass or ceramic enclosures, with plastic occasionally being used in low cost, low precision applications. The metal enclosures can be further subdivided into solder seal, resistance weld and cold weld types, the subdivision being made according to the technique used to seal the cover of the enclosure to the header or base.

Figure 5.2 shows the structure of typical solder seal, resistance weld, and cold weld bases. In the solder seal and cold weld cases, the lead throughs are brought through the header by matched glass-to-metal seals, whereas in the resistance weld case the seal is of the compression type. In all cases the actual

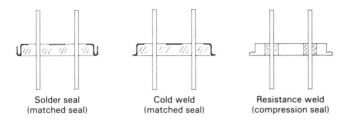

Fig. 5.2 Typical crystal headers.

mounting structure is typically welded to the lead throughs and can be chosen independently of the sealing method. Many different types of mount have been used over the years, and some of the most common are shown in Fig. 5.3. The choice of mount is largely determined by mechanical considerations, with rugged, stiff mounts being required to support blanks likely to be subjected to high shock levels, and more compliant mounts being required to minimize stresses on the crystal due to the mount itself. For large, low frequency blanks the problem of providing sufficient mechanical support without at the same time imposing undue stresses on the blank is particularly difficult, whereas on the other hand at higher frequencies the manufacturer has a relatively wide choice of satisfactory mount systems.

Whatever mount is chosen, the blank has to be secured in place with some form of adhesive, which must provide both mechanical strength and resilience and also be electrically conductive. The most commonly used are silver loaded epoxy resins or silver loaded polyimide pastes, the latter being more useful for higher temperature applications. The paste or resin is usually applied with a syringe or needle, and invariably requires to be cured at an elevated temperature before further processing. It is important that the paste does not out-gas appreciably during the curing cycle and so contaminate the crystal, and also important that the paste should not contain any chemical that will attack the electrode material and cause open-circuits. The latter can also be caused by paste that shrinks on curing. Figure 5.4 shows a selection of mounted and bonded blanks.

5.4 ADJUSTING TO FREQUENCY

After the final lapping stage, the tolerance on the blank frequency is of the order of 0.1%. After etching, the blank frequency is typically left higher than the desired nominal frequency by an amount of the order of 1% of the nominal frequency, allowing for the frequency to be brought down to nominal by the mass loading effect of the electrodes. The total amount by which the frequency is to be lowered in the base-plating and frequency

ELECTRICAL PROCESSING

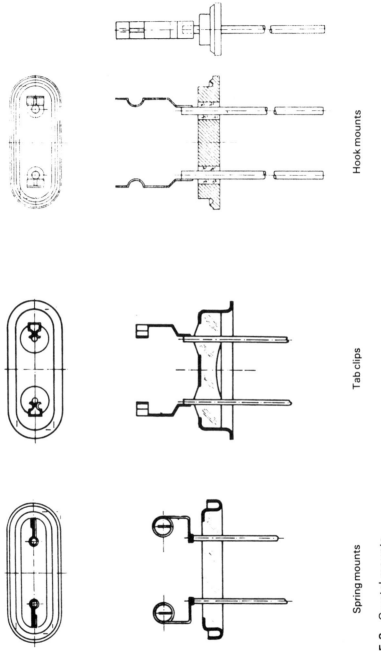

Fig. 5.3 Crystal mounts.

ADJUSTING TO FREQUENCY 83

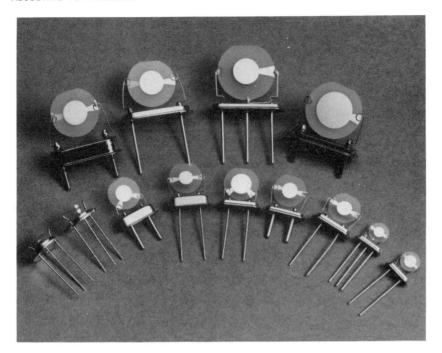

Fig. 5.4 Mounted crystal units.

adjusting processes is termed the *plateback* and is a key design factor. The final frequency tolerance required in the finished resonator is usually of the order of 10^{-5} or 10 ppm, that is two orders of magnitude tighter than the typical blank frequency tolerance after lapping and etching. Thus, although it would be very advantageous if the base-plating process were sufficiently accurately controlled to be able to plate to the nominal frequency, it is clear that the blank tolerances are generally too wide to allow this. Consequently, the objective in base-plating is to achieve a plateback of perhaps 80% of the total required, leaving the final frequency adjustment of the blanks to a later stage.

If this approach is followed, the crystal units after base-plating and mounting and bonding will be functional resonators with a frequency up to approximately 0.2% or 2000 ppm away from nominal. Each unit is then individually adjusted by evaporating small amounts of electrode material on to the resonator surfaces while observing the resonator frequency as measured by a suitable test set. Figure 5.5 shows an example of equipment designed for the automatic adjustment of a number of crystal units. The units are fitted with plating masks to limit the area to be plated to the central region of the base-plated electrodes, and then loaded into a carousel in a vacuum chamber. The chamber is evacuated and a stepping motor advances the carousel so that each crystal in turn is brought into the adjusting position. In this position the

Fig. 5.5 Automatic adjuster.

crystal is situated between two filaments and also linked in to a frequency measuring system. The crystal frequency is measured and compared to a preset target frequency. Assuming that the measured frequency is above the target frequency, silver is evaporated from the filaments on to the crystal, bringing down the frequency by the mass loading effect. The system monitors the crystal frequency, and interposes shutters between the crystal and the filaments when the target frequency is reached.

An alternative to this procedure is known as *direct plating*, and omits the base-plating stage altogether. In direct plating the crystal blanks are cleaned and etched, and then passed straight to the mounting and bonding operation. At this stage of course the crystals have no electrodes and so are not functional, but are nevertheless loaded into an equipment similar to that described above. The key difference now is that the plating masks used must define the full electrode pattern, including the electrode 'tails' which provide electrical connection to the mounts. As each crystal in turn is indexed into the adjusting position, the filaments are first of all switched on for a set time period to deposit sufficient electrode material for the film to become conducting and for the crystal to start operating. Once the crystal begins to oscillate, then the frequency can be monitored and the process continues as before.

Direct plating has some significant advantages in small-scale production and also in regard to the superior control of the motional parameters of the crystal unit that it affords. There are also corresponding disadvantages relating to large volume production, and some doubts concerning the precision of the adjustment process related to the observed drift in resonator frequencies after the completion of the process.

5.5 SEALING

As already mentioned in Section 5.3, apart from low cost, low precision units, the majority of resonators produced, are packaged in hermetically sealed metal, glass or ceramic packages. Glass holders can be of the tubular type, similar to thermionic valve enclosures, or more commonly of identical outline dimensions to the standard metal holders. The headers for these latter type of glass units generally contain a metal ring which in the sealing process is heated by RF induction techniques until the surrounding glass melts and fuses with the glass of the cover. The process is very suitable for use with high temperature, high vacuum bake-outs immediately before sealing, and can produce units with extremely good ageing rates, but in volume production is difficult to control and the pieceparts are relatively expensive. The recently introduced ceramic packages (Peters, 1976) require specialized sealing processes and are not yet in wide commercial use, so consequently the various types of metal enclosure are those most widely used at present.

Solder seal holders are, as the name implies, sealed by soldering the cover to the base or header. This is the oldest method of sealing metal holders, and suffers from the major disadvantage of contamination of the crystal by the flux used in the soldering operation. Solder sealing has consequently largely been replaced by resistance welding, which besides being free from flux problems, also lends itself to automation. Resistance welding depends on the fusing of the material of cover and header caused by the passage of a heavy current through the flanges of the pieceparts, which are held under moderate pressure in a dieset. Figure 5.6 shows a typical welder. The welding current can be either AC, or a DC pulse obtained by the discharge of a capacitor bank. The welding operation is normally carried out in a glovebox in a dry, inert atmosphere, the crystal units being brought in to the system via a vacuum oven and removed via an air-lock.

Cold welding is the term used to describe welds obtained by applying extreme pressure to the mating surfaces in such a way that any contaminated or oxidized surface layers are pushed aside to allow intimate contact between the atomically clean metals. This process, which requires precision tooling in order to ensure a uniform seal all around the flanges of the base and cover, has the advantages of not requiring any application of heat and of being

Fig. 5.6 Resistance welder.

easily built in to a high vacuum system. Hence cold weld units can be sealed under high vacuum or backfilled with inert gases without difficulty, whereas resistance weld units generally cannot be sealed under vacuum. Figure 5.7 shows a typical cold welding set up, incorporating a cryogenic pumping system and automatic control of the welding and vacuum cycle.

5.6 PHOTOLITHOGRAPHIC TECHNIQUES

The techniques described in this and the previous chapter are typical of those used in the processing of conventional resonators of the AT type. Similar techniques are used in the processing of doubly rotated resonators such as the SC-cut, although since these usually will be aimed at very high precision applications, the degree of process sophistication employed will generally be much higher than implied here. Different manufacturing techniques are required, however, for the new generation of miniature resonators such as the miniature GT-cut, tuning fork resonators, and miniature AT bar resonators.

The initial optical processing stages are essentially similar to the corresponding stages of conventional resonators, except that the emphasis is on the

Fig. 5.7 Cold weld press.

production of wafers that may measure one or two inches square, much larger than the conventional crystal blank. From the wafer stage onwards, the processing is much more akin to the production of SAW devices or semi-conductor devices than conventional resonators. The key process steps are the use of photolithographic methods, rather than the traditional shadow masks, to define the metallization patterns on the wafer surfaces, and the use of deep etching to separate the individual resonators. More details and further references can be found in the survey by Moore (1983).

Part 3
The crystal as a circuit element

6 Equivalent circuit analysis

6.1 EQUIVALENT CIRCUITS

Figure 6.1 shows the conventionally accepted equivalent circuit of a crystal resonator at a frequency near its main mode of vibration. The circuit elements L_1, C_1, R_1 are the electrical equivalents of the inertia, stiffness, and internal losses of the mechanical vibrating system. If the crystal were clamped in such a way that no vibration were possible, this arm would be absent, and hence L_1, C_1 and R_1 are known as the *motional parameters* of the crystal. The element C_0 represents the capacitance of the capacitor formed by the electrodes of the crystal and the quartz dielectric. It can be measured as the effective capacitance of the crystal unit at frequencies far removed from resonance, and is known as the *static capacitance*.

One of the principal characteristics of a quartz resonator as compared to *LC* circuits or other types of mechanical resonator, is that the Q factor of the motional arm is extremely high. For commercially available units, typical values range from 20 000 to several hundred thousand, while specially designed units can have Q values of several million. These compare with Qs of less than 1000 for the best wound inductors.

The *motional capacitance* C_1 for typical AT-cut fundamental mode resonators is typically in the range 10 to 30 fF (1 fF = 10^{-15} F). For overtone resonators, C_1 reduces in inverse proportion to the square of the overtone order, so that for a third overtone typical values range from 1 to 3 fF.

As indicated above, the static capacitance C_0 is essentially determined by the electrode size and separation and is thus independent of the overtone order. Typical values for AT-cuts range from 1 to 7 pF. More important in

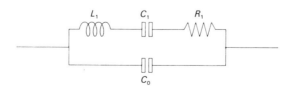

Fig. 6.1 Equivalent circuit of crystal resonator.

practice is the ratio of C_0 to C_1 because this ratio effectively determines the sensitivity of the crystal frequency to changes in external circuit parameters. As shown in Chapter 2, the capacitance ratio is intimately connected with the electromechanical coupling factor for the particular crystal cut in question. For fundamental mode ATs above about 10 MHz, the ratio is approximately 200, whereas for overtones it increases roughly in proportion to the square of the overtone order. For low frequency fundamentals where the crystal blank is partially or fully contoured, C_0/C_1 again increases because the contour limits C_1 but not C_0. The 'pulling' sensitivity is thus greatest for the higher frequency fundamental mode units.

6.2 LOSSLESS CIRCUIT ANALYSIS

6.2.1 Characteristic frequencies

Precise analysis of the equivalent circuit reveals several characteristic frequencies. In most practical cases, because of the very high resonator Q, it is sufficient to consider only two. These are the *series resonance* frequency f_r and the *anti-resonance* or *parallel* resonance frequency f_a. These correspond to the *natural modes* of the resonator under short-circuit and open-circuit conditions, respectively. Expressions for these frequencies can easily be derived by analysis of the lossless circuit obtained by neglecting the motional resistance R_1 as shown in Fig. 6.2.

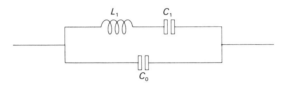

Fig. 6.2 Lossless equivalent circuit.

If Y is the admittance of the crystal unit, then Y can be written as

$$Y = jB_0 + jB_1$$

where $B_0 = \omega C_0$, ω is the angular frequency, and jB_1 is the admittance of the motional arm. jB_1 is given by the reciprocal of the motional arm impedance, ie, by

$$jB_1 = 1/(j\omega L_1 + 1/j\omega C_1) = j\omega C_1/(1 - \omega^2 L_1 C_1)$$

Therefore

$$Y = j(B_0 + B_1) = j\omega C_0 (1 + C_1/C_0 - \omega^2 L_1 C_1)/(1 - \omega^2 L_1 C_1)$$

LOSSLESS CIRCUIT ANALYSIS

and if Z is the crystal impedance,

$$Z = 1/Y = (1 - \omega^2 L_1 C_1)/\{j\omega C_0(1 + C_1/C_0 - \omega^2 L_1 C_1)\}$$

The series resonance frequency f_r corresponds to the zero of the impedance Z, and the anti-resonance frequency f_a corresponds to the zero of Y. Consequently, if ω_r and ω_a are the corresponding angular frequencies, then

$$\omega_r^2 = 1/(L_1 C_1)$$

and

$$\omega_a^2 = (1 + C_1/C_0)/(L_1 C_1)$$

or

$$\omega_a^2 = \omega_r^2(1 + C_1/C_0)$$

Since $C_1/C_0 \ll 1$, to a good degree of approximation the last equation can be written

$$\omega_a = \omega_r(1 + C_1/2C_0)$$

or

$$(\omega_a - \omega_r)/\omega_r = (f_a - f_r)/f_r = C_1/2C_0 \qquad (6.1)$$

which directly relates the separation between the resonance frequencies to the capacitance ratio.

6.2.2 Crystal with a load capacitance

Many practical oscillator circuits make use of a load capacitor C_L in series or parallel with the crystal, either in order to provide a means for final frequency adjustment, or perhaps for modulation or temperature compensation purposes. The presence of the load capacitor shifts the working frequency of the crystal by an amount depending upon the value of C_L and the values of C_0 and C_1. Figures 6.3 and 6.4 show the series and parallel connections respectively. Figures 6.5 and 6.6 show plots of the impedance and admittance

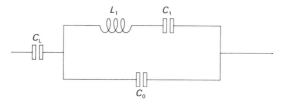

Fig. 6.3 Crystal with series load capacitor.

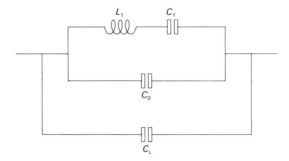

Fig. 6.4 Crystal with parallel load capacitor.

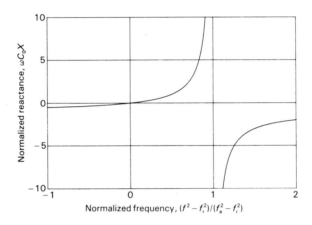

Fig. 6.5 Normalized crystal reactance vs frequency.

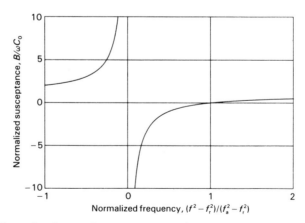

Fig. 6.6 Normalized crystal susceptance vs frequency.

of a crystal unit in the lossless approximation. From these plots it is clear that when C_L is connected in parallel with the crystal, the resonance frequency f_r is not affected, but that the anti-resonance frequency f_a is shifted down to a frequency f_L. Similarly, when C_L is connected in series with the crystal, f_a is unaffected but f_r is shifted up to a frequency f_L. It follows from the physical interpretation of f_r and f_a as short- and open-circuit resonances, respectively, that f_L has the same value in both the above cases, since an open-circuit parallel combination is the same as a short-circuit series combination. The value of f_L can therefore be very easily obtained from the expressions for f_a already given by substituting f_L for f_a and $(C_0 + C_L)$ for C_0. Therefore

$$f_L^2 = f_r^2(1 + C_1/(C_0 + C_L))$$

or

$$(f_L - f_r)/f_r = C_1/2(C_0 + C_L) \tag{6.2}$$

This is the *load resonance frequency offset*.

The sensitivity of the working frequency to small changes in C_L, ie, the 'pulling sensitivity', is given by

$$df_L/dC_L = -C_1/2(C_0 + C_L)^2$$

For high overtone orders, the sensitivity is orders of magnitude lower than for fundamental mode units, because of the much lower typical C_1 values and consequently high ratio of C_0 to C_1.

From Eqns (6.1) and (6.2) it follows that

$$(f_L - f_r)/(f_a - f_r) = C_0/(C_0 + C_L) = k$$

For a purely capacitive load where C_L varies from zero upwards, the parameter k ranges from 1 to 0. This corresponds to the fact that the working frequency f_L is restricted to the range between f_r and f_a. Thus the maximum load resonance frequency offset is determined through Eqn (6.1) by the capacitance ratio C_0/C_1, and the location of f_L within this range is determined by k, ie, the value of C_L as compared to C_0. For example, if $C_L = 4C_0$, then $k = 0.2$ and the working frequency is shifted relative to f_r by 20% of the pole-zero spacing $(f_a - f_r)$.

6.3 LOSSY CIRCUIT ANALYSIS

The simple analysis of the characteristic frequencies of a crystal resonator based on the lossless equivalent circuit glosses over the subtle differences between the several 'resonance' frequencies that may be defined when losses are present. The existence of these several frequencies can easily lead to confusion in comparing the results of measurements made by different methods, especially at high frequencies and overtones so that an analysis of the equivalent circuit including losses is necessary.

6.3.1 Motional impedance and admittance

At an angular frequency ω the motional impedance Z_1 can be written

$$Z_1 = R_1 + j\omega L_1 + 1/j\omega C_1$$

or

$$Z_1 = R_1 + jL_1(\omega^2 - \omega_s^2)/\omega$$

where $\omega_s^2 L_1 C_1 = 1$ and ω_s is the angular frequency corresponding to the motional resonance frequency f_s where Z_1 is purely resistive.

If now the normalized frequency variable x is introduced by the relation

$$x = (f^2 - f_s^2)/(f_p^2 - f_s^2) = (\omega^2 - \omega_s^2)/(\omega_p^2 - \omega_s^2)$$

where $f_p^2 = f_s^2(1 + C_1/C_0)$ is the frequency corresponding to the anti-resonance frequency in the lossless case, then

$$Z_1 = R_1 + jX_1$$

where

$$X_1 = x(\omega_p^2 - \omega_s^2)L_1/\omega$$

From the definition of ω_p and ω_s, it follows that

$$(\omega_p^2 - \omega_s^2)L_1 = \omega_s^2 L_1 C_1/C_0 = 1/C_0$$

Therefore $X_1 = x/\omega C_0 = xX_0$ where $X_0 = 1/\omega C_0$. To a good approximation, X_0 can be regarded as a constant with a value equal to the magnitude of the reactance of C_0 at frequency ω_s, ie, $X_0 \sim 1/\omega_s C_0$.

It is convenient at this point to introduce the parameters Q, r, and M. Q is defined as the *quality factor* of the motional arm, ie, $Q = \omega_s L_1/R_1$, r is the capacitance ratio C_0/C_1, and M is the *figure of merit* Q/r. From the definition, it follows that $M = X_0/R_1$. Then the expression for Z_1 can be written in normalized form as

$$Z_1/R_1 = 1 + jxM$$

The motional admittance Y_1 is the reciprocal of the motional impedance. Thus if y_1 is the normalized admittance $R_1 Y_1$, it follows immediately that

$$y_1 = 1/(1 + jxM) = (1 - jxM)/(1 + x^2 M^2)$$

If $y_1 = g_1 + jb_1$, then

$$g_1 = 1/(1 + x^2 M^2)$$
$$b_1 = -xM/(1 + x^2 M^2)$$

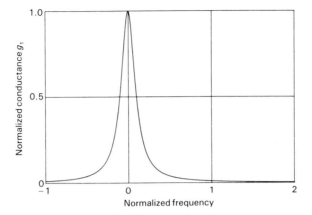

Fig. 6.7 Normalized crystal conductance vs frequency ($M = 10$).

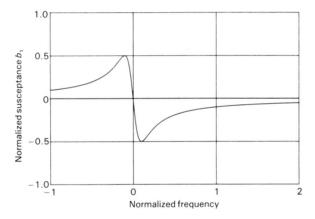

Fig. 6.8 Normalized motional susceptance ($M = 10$).

Figures 6.7 and 6.8 show plots of g_1 and b_1 as functions of the normalized frequency x. g_1 has a single maximum of value 1 at $x = 0$ and is always > 0. b_1 is > 0 for $x < 0$ and < 0 for $x > 0$, and has extrema at the points $x = \pm M^{-1}$. The extreme values are $+\frac{1}{2}$ at $x = -M^{-1}$ and $-\frac{1}{2}$ at $x = M^{-1}$.

The magnitude squared of the normalized admittance is $g_1^2 + b_1^2$ and is given by

$$g_1^2 + b_1^2 = 1/(1 + x^2 M^2)$$

and again has a maximum of 1 at $x = 0$.

6.3.2 Crystal admittance

The crystal admittance Y is the sum of the motional admittance and the admittance of the shunt capacitance C_0. In the present degree of approximation, the admittance of C_0 is treated as a constant of value jB_0 with $B_0 = \omega_s C_0 = 1/X_0$. Its normalized value is thus $jR_1/X_0 = jM^{-1}$. Hence if the normalized crystal admittance is $y = g + jb$,

$$g = g_1 = 1/(1+x^2M^2)$$
$$b = M^{-1} - xM/(1+x^2M^2)$$

Clearly, the real part of the crystal admittance is just the real part of the motional admittance, being always >0 and having a single maximum of normalized value 1 at $x=0$. The imaginary part is shifted by the additional contribution of C_0. As a consequence, the susceptance b may have two, one, or no zeros rather than the single zero at $x=0$ possessed by the motional susceptance. The positions of the zeros on the x axis are given by the roots of the quadratic equation

$$x^2 - x + M^{-2} = 0$$

ie,

$$x = \{1 - (1-4M^{-2})^{1/2}\}/2$$

or

$$x = \{1 + (1-4M^{-2})^{1/2}\}/2$$

If $1 > 4M^{-2}$ or alternatively if $M > 2$, then two zeros exist. If $M=2$ then the zeros coincide, and if $M<2$ then there are no zeros.

For the case $M \gg 2$, the two zero locations are accurately given by the approximate formulae

$$x = M^{-2}$$
$$x = 1 - M^{-2}$$

On the other hand, when $M=2$, the zeros coincide at

$$x = 0.5$$

Figure 6.9 shows a normalized plot of the crystal susceptance as a function of the normalized frequency x. In the present degree of approximation where C_0 is regarded as having a constant admittance in the frequency range of interest, the extrema of b occur at the same frequencies as those of b_1, ie, at $x = \pm M^{-1}$.

NOTE: the assumption that C_0 can be regarded as having a constant reactance or admittance in the frequency range of interest is well justified in the case of quartz resonators where the resonance frequencies are very closely grouped. However, for resonators made from materials with large

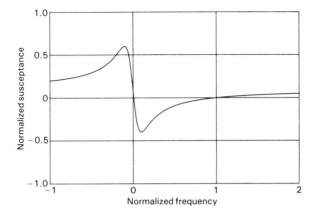

Fig. 6.9 Normalized crystal susceptance ($M = 10$).

electromechanical coupling and hence widely spaced resonance frequencies, the assumption may not be justified. Then an exact analysis such as that of Ballato (1970) must be used.

6.3.3 Phase characteristics

The phase angle ϕ of the crystal admittance is given by

$$\tan(\phi) = b/g$$
$$= (1 + x^2 M^2 - xM^2)/M$$

Of particular importance in the zero phase measurement technique (Chapter 8), and in frequency stability considerations (Chapter 9), is the slope of the phase-frequency characteristic in the region of zero phase. Differentiating with respect to the normalized frequency x leads at once to

$$d(\tan(\phi))/dx = M(2x - 1)$$

Differentiating the defining equation for x with respect to the angular frequency ω gives

$$dx/d\omega = 2\omega Q R_1 C_0/\omega_s = 2Q/(\omega_s M)$$

where the last expression is valid to a high degree of accuracy in the vicinity of the resonance frequencies. Combining these expressions, it follows that

$$d(\tan(\phi))/d\omega = 2(Q/\omega_s)(2x - 1)$$

and also that

$$d\phi/d\omega = 2(Q/\omega_s)(2x - 1)\cos^2(\phi)$$

Clearly then in the neighbourhood of the series resonance frequency where x and ϕ are close to zero, the phase slope is primarily determined by the resonator Q.

6.3.4 The admittance circle

It can easily be verified from the expressions for g and b given previously, that

$$(g - \tfrac{1}{2})^2 + (b - M^{-1})^2 = (\tfrac{1}{2})^2$$

This is the equation of a circle in the $g + jb$ plane, with radius $1/2$ and centre at the point $(\tfrac{1}{2}, M^{-1})$. Figure 6.10 illustrates the case for $M = 10$. Each point on the circle corresponds to a value of frequency, with the frequency increasing as the circle is traversed in a clockwise direction.

The zeros of b, calculated in the preceding section, appear in Fig. 6.10 as the intersections of the admittance circle with the g axis. Clearly, as M decreases, the circle moves upwards along the b axis. The points of intersection gradually move closer until at $M = 2$ they coincide at the point $(\tfrac{1}{2}, 0)$. For M smaller than 2, the circle does not intersect the g axis.

The point P_2 in Fig. 6.10 is the point where the magnitude of the crystal admittance reaches its maximum. It lies on the line from the origin of the g, b plane through the centre of the admittance circle. It thus follows that at the frequency of maximum admittance

$$b/g = M^{-1}/0.5 = 2/M$$

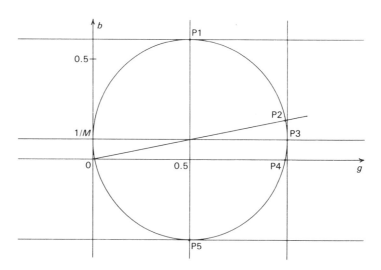

Fig. 6.10 Admittance circle ($M = 10$).

The frequency of maximum admittance is therefore a root of the quadratic equation

$$x^2 - x - M^{-2} = 0$$

The roots are

$$x = \{1 - (1+4M^{-2})^{1/2}\}/2$$
$$x = \{1 + (1+4M^{-2})^{1/2}\}/2$$

For $M \gg 1$, the roots are accurately given by

$$x = -M^{-2}$$
$$x = 1 + M^{-2}$$

The root $x = -M^{-2}$ is the frequency of maximum admittance, and the other root is the frequency of minimum admittance.

6.3.5 The resonance frequencies

The admittance circle in Fig. 6.10 shows very closely the distinction between the three characteristic frequencies which may be loosely termed resonance frequencies of the crystal. The frequencies in question are (a) the frequency of maximum admittance, f_m, (b) the frequency of maximum conductance, f_s, and (c) the frequency of zero susceptance or zero phase, f_r. These correspond to the points P_2, P_3, and P_4, respectively, in Fig. 6.10. In addition, there is another pair of characteristic frequencies close to the normalized frequency $x = 1$. These are the frequency of minimum admittance, f_n, and a second frequency of zero susceptance or phase, f_a. However, the first group is of most practical importance:

(a) The frequency of maximum admittance f_m is that frequency that would be measured in a simple transmission test circuit, and corresponds to the frequency at which maximum current flows through the crystal.
(b) The frequency of maximum conductance f_s is the resonance frequency of the motional arm of the equivalent circuit. It is the frequency as measured by admittance bridge methods when the capacitance C_0 is balanced out, or by transmission methods when C_0 is either physically tuned out or compensated for mathematically.
(c) The frequency of zero phase f_r, if it exists, is the lower of the two frequencies at which the crystal presents a purely resistive impedance. It has been adopted by the IEC in Publication 444 as the standard parameter for the characterization of crystal frequency.

6.3.6 Numerical estimates for the differences between f_m, f_s, and f_r

If the normalized frequencies corresponding to $f_m, f_s,$ and f_r are $x_m, x_s,$ and x_r, respectively, then from the preceding analysis

$$x_m = \{1 - (1+4M^{-2})^{1/2}\}/2$$
$$x_s = 0$$
$$x_r = \{1 - (1-4M^{-2})^{1/2}\}/2$$

Figure 6.11 shows x_m and x_r as a function of the figure of merit M for the range $M = 2$ to $M = 100$. Clearly for $M > 10$ both x_m and x_r are both close to zero, allowing the use of the approximate expressions

$$x_m = -1/M^{-2}$$
$$x_r = +1/M^{-2}$$

Removing the frequency normalization leads to the following expressions for the fractional frequency differences

$$(f_m - f_s)/f_s = -r/(2Q^2)$$
$$(f_r - f_s)/f_s = +r/(2Q^2)$$

For AT fundamental mode crystals with $r \sim 200$ and $Q \sim 50\,000$ the fractional frequency error is 4×10^{-8} or 0.04 ppm, which is practically negligible.

For third overtone resonators with $r \sim 2000$, and $Q \sim 100\,000$ the error increases to 1×10^{-7} or 0.1 ppm, still negligible for most purposes.

However, for fifth and higher overtones the frequency differences become significant. For a fifth overtone unit with $r \sim 7000$, and $Q \sim 70\,000$, the error is 0.7 ppm, whereas for seventh and ninth overtone units the figure of merit will generally be sufficiently low to warrant the use of the exact

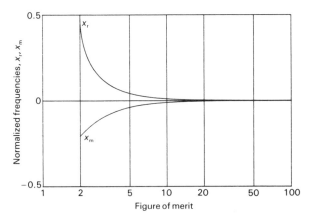

Fig. 6.11 Normalized frequencies vs figure of merit.

formulae for the frequency error. Taking an $r \sim 20\,000$ and $Q \sim 60\,000$ as typical of a seventh overtone crystal, then $M = 3$ and

$$x_m = -0.1$$
$$x_r = +0.127$$

Therefore the fractional frequency errors are -2.5 ppm and $+3.2$ ppm, respectively.

For a ninth overtone, the figure of merit M will typically be <2, so that f_r will not exist. Assuming $M = 1.5$, x_m will be -0.33 and for an r value of 40 000, the fractional frequency error of f_m will be -4.2 ppm.

6.3.7 Crystal impedance at resonance and anti-resonance

For a crystal with a figure of merit $M > 10$, it has been demonstrated that the resonance frequencies are closely grouped around the normalized frequency $x = 0$, and the anti-resonance frequencies are correspondingly grouped around $x = 1$. At these frequencies, the crystal susceptance and therefore the reactance are close to zero, so that the normalized crystal impedance z is almost purely resistive and is given by $z = 1/g$.

If r_r and r_a are the impedance values at resonance and anti-resonance, ie, at $x = 0$ and $x = 1$, then from the expression

$$g = 1/(1 + x^2 M^2)$$

it follows that

$$r_r = 1$$
$$r_a = 1 + M^2 \sim M^2$$

NOTE: in the older literature, the term EPR or 'equivalent parallel resistance' is often used for the resistance of a crystal unit when operated at parallel or anti-resonance with a parallel load capacitor C_L. Removing the normalization in the expression for r_a above and writing $C_0 + C_L$ for C_0 gives the following relation between the EPR and the motional resistance R_1:

$$\text{EPR} = 1/[R_1(\omega_s(C_0 + C_L))^2]$$

6.4 EFFECTIVE CRYSTAL PARAMETERS WITH A SERIES LOAD CAPACITOR

The shift in the resonance frequencies of a crystal unit caused by a load capacitor has already been discussed in Section 6.2.1 on the basis of the lossless equivalent circuit. Analysis including losses demonstrates that a crystal unit with a series load capacitor is equivalent to a crystal unit without a load capacitor but with modified parameter values. The equivalence can be used to provide a means of obtaining parameter values that would otherwise be impracticable. This is particularly useful in filter design where a wide range of motional inductance values is required.

6.4.1 Circuit equivalence

Figure 6.12 shows the equivalent circuit of a resonator with a series load capacitor C_L. Figure 6.13 shows the equivalent circuit of a resonator with motional parameters L_1', C_1', and R_1', and static capacitance C_0'. The circuits of Figs. 6.12 and 6.13 are equivalent provided that the element values of Fig. 6.13 are suitably defined in terms of those of Fig. 6.12.

First, consider the behaviour of the circuits at high frequencies. The inductive arms may then be regarded as open-circuit, when it becomes clear that a necessary condition for equivalence is that

$$1/C_0' = 1/C_0 + 1/C_L$$

ie

$$C_0' = C_0 C_L / (C_0 + C_L)$$

For an arbitrary angular frequency ω, it then follows that

$$Z_0' = Z_0 + Z_L$$

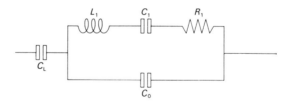

Fig. 6.12 Crystal with series load capacitor.

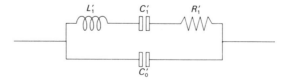

Fig. 6.13 Equivalent circuit for Fig. 6.12.

where Z_0', Z_0, and Z_L are the impedances of C_0', C_0, and C_L at frequency ω. Defining Z_m and Z_m' to be the impedances of the motional arms in Figs. 6.12 and 6.13, respectively, then by inspection the total admittance Y of crystal and load capacitor in Fig. 6.12 can be written

$$Y = (Z_0 + Z_m)/(Z_0 Z_L + Z_L Z_m + Z_m Z_0)$$

If now the admittance $j\omega C_0'$ is subtracted from Y, the equivalence is proved if the resulting impedance $1/(Y - j\omega C_0')$ can be put into the form Z_m'.

$$Y - j\omega C_0' = (Z_0 + Z_m)/(Z_0 Z_L + Z_L Z_m + Z_m Z_0) - 1/(Z_0 + Z_L)$$
$$= Z_0^2/(Z_0 + Z_L)(Z_0 Z_L + Z_L Z_m + Z_m Z_0)$$

Therefore

$$1/(Y - j\omega C_0') = (1 + Z_L/Z_0)\{Z_L + (1 + Z_L/Z_0)Z_m\}$$

Since $Z_L/Z_0 = C_0/C_L$, a real constant K can be defined by

$$K = 1 + Z_L/Z_0 = 1 + C_0/C_L$$

Clearly, K is always > 1, since C_0 and C_L are positive. Then

$$1/(Y - j\omega C_0') = K(Z_L + K Z_m)$$

which has the form of Z_m' provided that

$$L_1'/L_1 = R_1'/R_1 = K^2$$
$$1/C_1' = K/C_L + K^2/C_1$$

Both the motional resistance and inductance are multiplied by the factor K^2. Since $K > 1$ for all values of C_L, the effective parameters are always increased by the addition of a series load capacitor. Since both resistance and inductance are increased by the same factor, the Q factor remains unchanged. The effective motional capacitance is however always decreased.

The effective capacitance ratio C_0'/C_1' is easily calculated to be

$$C_0'/C_1' = (C_0/C_1)\{(C_1 + C_0 + C_L)/C_L\}$$

or to a close degree of approximation

$$C_0'/C_1' = K(C_0/C_1)$$

Thus the effective capacitance ratio is always increased. Since the figure of merit M is defined as the quotient Q/r, where r is the capacitance ratio, and since the effective Q is not changed by the addition of a load capacitance, it follows that the presence of a load capacitance always degrades the figure of merit.

From the formulae given above for L_1' and C_1', it easily follows that if ω_s and ω_s' are defined by $\omega_s'^2 L_1' C_1' = \omega_s^2 L_1 C_1 = 1$ then

$$\omega_s'^2 = \omega_s^2 \{(1 + C_1/(C_0 + C_L)\}$$

To a good degree of approximation

$$(\omega_s' - \omega_s)/\omega_s = C_1/2(C_0 + C_L)$$

In the lossless case dealt with in Section 6.2.2, these formulae give directly the shift of the load resonance frequency relative to the resonance frequency. In the general case, for most situations of practical interest, ie, where the figure of merit is >10, the resonance frequency is still very close to the frequency ω_s, so that the formulae can still be applied with little error.

In the lossless case, the anti-resonance frequency ω_a is equal to the frequency ω_p defined by

$$\omega_p^2 = \omega_s^2(1 + C_1/C_0)$$

It is easily shown that if the frequency ω_p' is defined by

$$\omega_p'^2 = \omega_s'^2(1 + C_1'/C_0')$$

then $\omega_p' = \omega_p$. Thus the lossless anti-resonance frequency is not shifted by a series load capacitor. In the lossy case, provided the figure of merit is sufficiently large, the anti-resonance frequency remains close to ω_p, so that for practical purposes the shift in the anti-resonance frequency can be ignored.

It is sometimes the case that a crystal filter design requires a wider range of crystal inductances than can easily be achieved directly. The equivalence of the circuits in Figs. 6.12 and 6.13 provides one solution to the solution by allowing the use of a low inductance crystal with a series capacitor to simulate a higher inductance crystal.

To state the problem precisely, assume that the filter design requires a crystal of frequency f_s' and motional inductance L_1', while the available crystals have a motional inductance of L_1 and a static capacitance C_0. The problem is to determine a load capacitance C_L and a frequency f_s such that a crystal with parameters f_s, L_1 and C_0, in series with C_L, has the effective parameters f_s' and L_1'. The step-by-step procedure is

(a) Calculate $K = 1 + C_0/C_L$ from $K^2 = L_1'/L_1$.
(b) Calculate C_L from K.
(c) Calculate f_s from $(f_s' - f_s)/f_s' \sim C_1/2(C_0 + C_L)$ where C_1 is obtained from $\omega_s'^2 L_1 C_1 \sim 1$.

7 Characteristics of AT-cut crystal resonators

7.1 BLANK DIAMETER AND GEOMETRY

The conventional AT-cut resonator, as opposed to the newer miniature AT bar resonators, is in the first approximation a pure thickness mode resonator (Chapter 2). For circular plates of sufficiently large diameter to thickness ratio d/t this first approximation provides a reasonable guide to resonator behaviour, but as the ratio decreases the effects of the finite lateral dimensions become increasingly important (Chapter 3).

For d/t ratios greater than 50, a flat or plano-plano blank design with no bevel or contour will usually provide acceptable performance for fundamental mode crystals. For d/t ratios less than 50 but greater than 30, an edge bevel (partial contour) will generally be necessary, whereas for ratios below 30 a complete spherical contour is required. Provided that d/t remains above 12 a plano-convex design is usually adequate, but for the lowest d/t ratios a full bi-convex contour is required with perhaps a steeper edge bevel in addition. The limiting values of the d/t ratios quoted above are empirical in nature and by no means absolute, particularly in the lower d/t ranges, where for example it might often prove desirable to use a plano-convex design with an edge bevel rather than a bi-convex design to achieve some desired combination of parameters. Nevertheless, the classification given is useful in estimating the probable complexity required in a given blank design. For overtone mode resonators, the limiting values can usually be relaxed as compared to fundamentals so that flat designs can safely be used at d/t ratios below 50, perhaps down to $d/t = 35$. It is, however, difficult to give general rules for contoured overtone resonators since these are little used in general applications, mostly being restricted to limited volume, high precision units at specific frequencies.

When these design rules are combined with information on the maximum blank diameters that can be accommodated in a given holder style, it is straightforward to determine the frequency ranges for that holder style in which bevelled, plano-convex and bi-convex designs need to be used. This is

108 CHARACTERISTICS OF AT-CUT CRYSTAL RESONATORS

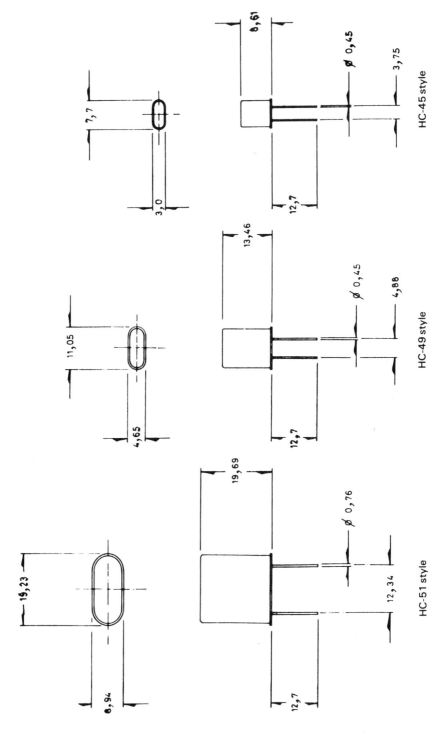

Fig. 7.1 Crystal holders.

BLANK DIAMETER AND GEOMETRY 109

important both in regard to the obvious differences in the manufacturing costs of say a bi-convex unit as compared to a flat unit, but also in regard to the different performance characteristics that will result in the various frequency intervals.

The maximum diameter blank that can be accommodated in a given holder is determined by the space available. Figures 7.1 and 7.2 give the outline dimensions of some of the more popular crystal holders, and Table 7.1 gives the approximate maximum blank diameters and the frequency ranges applicable to the different types of blank geometry. Again the data in Table 7.1 should be regarded as a guide only, manufacturing practices differing from supplier to supplier. The maximum blank diameter listed in Table 7.1 is a *maximum*, and will not generally be used for all units in a given holder. It is clearly desirable from the manufacturing point of view to use flat blanks, but when the frequency required is such that the necessary d/t ratio can be

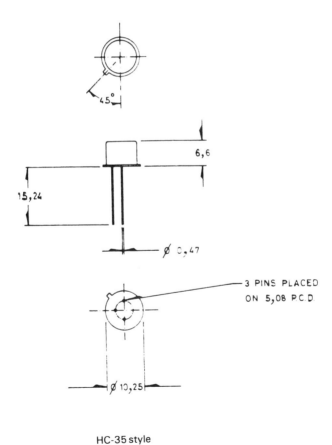

HC-35 style

Fig. 7.2 Crystal holders.

Table 7.1

	Max blank	Frequency ranges (MHz)			
Holder	diameter (mm)	Flat	Bevelled	Plano-convex	Bi-convex
HC-45	5.0	16 min	10 - 16	below 10	—
HC-35	6.2	13 min	8 - 16	below 8	—
HC-49	8.6	10 min	5.5 - 10	below 5.5	—
HC-51	15	5.5 min	3 - 5.5	1.5 to 3	below 1.5

achieved with blanks smaller than the maximum, then such blanks will generally be used. This results in economy in raw material, and also in mechanically more robust units, since very thin blanks of unnecessarily large diameter are prone to breakage.

7.2 MOTIONAL CAPACITANCE AND SHUNT CAPACITANCE

The shunt capacitance or C_0 is essentially the sum of the capacitance C_e between the electrodes of the crystal and the capacitance C_h of the holder itself. For flat blanks, neglecting fringing effects and the capacitance due to the electrode tails, the capacitance C_e in pF is approximately given by

$$C_e = 0.04A/t \qquad (7.1)$$

where A is the electrode area in mm² and t the blank thickness in mm. This can be recast in the form

$$C_e = 0.019d^2f/N \qquad (7.2)$$

where now d is the electrode diameter in mm, f the nominal frequency in MHz and N the overtone order. To obtain C_0, the holder capacitance C_h which is typically of the order of or less than 1 pF must be added to C_e as calculated from Eqn (7.1) or (7.2).

The C_0 of contoured crystals cannot be calculated directly from these simple formulae, but nonetheless Eqn (7.2) will give a reasonable estimate, sufficient for most practical purposes.

The motional capacitance of an AT-cut resonator can in principle be calculated from the second of Eqns (2.51), with appropriate values of the material constants. However, this is only valid for pure thickness modes and in practice needs correction to take account of the variation in the amplitude of vibration across the surface of the plate. Empirically it is found that at least for fundamental mode resonators, C_1 is given by the approximate formula

$$C_1 = 0.105d^2f/N^3 \qquad (7.3)$$

which differs from the formula obtained directly from Eqn (2.51) only in the value of the numerical factor.

Equation (7.3) predicts C_1 for fundamental resonators reasonably accurately, but for overtone resonators will only give satisfactory values for units, such as filter crystals, designed according to energy trapping principles. Standard overtone oscillator crystals are usually fabricated with mass loadings far in excess of that required by energy trapping theory (Chapter 3), and have C_1 values substantially lower than predicted by Eqn (7.3). Another cause of variation in the C_1 of overtone oscillator crystals lies in the commonly used method of final frequency adjustment by evaporation of material on to the central region of the electrodes. If too much material is added in the adjustment process, the result will be that the motional parameters of the unit begin to be governed by the diameter of the adjusting spot, rather than the diameter of the electrodes as originally deposited, because of energy trapping under the spot. Hence the conclusion can be drawn that the control of the motional capacitance of overtone resonators is substantially more difficult than for fundamentals.

The behaviour of the motional capacitance of contoured crystals is quite different from that of flat crystals. In the latter case, as indicated above, the electrode diameter is the main governing factor. With contoured crystals, however, the energy trapping due to the contour itself limits the vibrating area of the blank, and an effective diameter d_v can be defined in such a way that the amplitude of vibration can be regarded as negligible outside a circle of diameter d_v. Then it follows that however much the electrode diameter is increased over and above the value d_v, the motional capacitance will remain constant. In contrast, for electrode diameters less than d_v the C_1 decreases with electrode area much as for a flat crystal. In practice it is usual for the electrode diameter to be chosen some 20% larger than d_v so that the motional parameters are controlled by the blank geometry; if, as in the case of some filter crystals, it is necessary to use smaller electrode diameters, then another cause of variation in C_1 is the orientation of the electrode tails relative to the crystallographic X axis.

Equations (7.2) and (7.3) show that for flat crystals the capacitance ratio $r = C_0/C_1$ has the approximate value 180 N^2. This does not take into account the holder capacitance, so that in practice the minimum r value observed for fundamentals is about 200, with larger values for smaller electrode diameters when the holder capacitance is a larger part of the total C_0. Because of the departures of the C_1 value for overtone crystals from the predicted figures, the capacitance ratio for overtones is generally substantially greater than 180 N^2, the discrepancy increasing with the overtone. Likewise for contoured crystals, the capacitance ratio is higher than for flat crystals because of the limitation of the C_1 value by the contour already discussed.

Table 7.2 shows some typical values of the motional and shunt capacitance and the capacitance ratio of both contoured and uncontoured fundamental

Table 7.2 Motional and shunt capacitances

Frequency and overtone	C_1 (fF)	C_0 (pF)	(C_0/C_1)
5 MHz plano-convex fundamental	7	3.2	457
15 MHz plano-plano fundamental	25	5.5	220
45 MHz third overtone	2	5.5	2 750
75 MHz fifth overtone	0.6	5.5	9 170
105 MHz seventh overtone	0.15	3.5	23 300
135 MHz ninth overtone	0.08	3.5	43 750

mode resonators, and of overtone units up to the ninth overtone. Figure 7.3 shows the dependence of the C_1 and C_0 of a typical contoured blank on the electrode diameter, and the variation of the capacitance ratio. The electrode diameter corresponding to the minimum in the capacitance ratio curve is in a sense the optimum value, corresponding as it does to a maximum in the pulling sensitivity of the crystal unit.

7.3 Q FACTOR AND TIME CONSTANT

The Q of a resonator is inversely proportional to the ratio of the energy dissipated per cycle to the total stored energy. The energy dissipated can be assigned to several different causes, the chief of which are:

(1) Intrinsic acoustic losses in the body of the material, determined by the

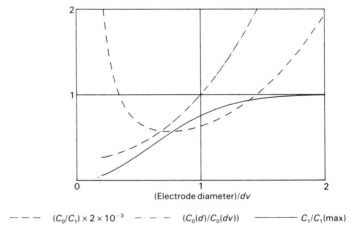

Fig. 7.3 Variation of C_1, C_0 and C_0/C_1 of a contoured crystal with electrode diameter.

basic physical properties of the material and the defects introduced by the growth process.
(2) Acoustic losses at the surface of the resonator due to the imperfections produced by the lapping operations.
(3) Electrical and acoustic losses in the electrode material, dependent on the material used, its conductivity and thickness, and film defects owing to poor metallization.
(4) Losses due to mounting which are determined by the resonator design and mounting techniques. The critical parameter is the diameter to thickness ratio.
(5) Acoustic losses due to atmospheric damping, which can be eliminated by sealing the units under vacuum.

If the Q factors due to each of these mechanisms alone are denoted by Q_k with k running from 1 to 5, then the overall Q is given by

$$Q^{-1} = Q_1^{-1} + Q_2^{-1} + Q_3^{-1} + Q_4^{-1} + Q_5^{-1} \qquad (7.4)$$

Q_1, the Q factor associated with the intrinsic acoustic loss of quartz can be regarded as a limiting value which cannot be exceeded by any improvements in design or processing. Following the discussion in Section 2.4, the intrinsic Q for the slow shear mode in an AT cut resonator is determined by the time constant τ for that mode, which is in turn determined in terms of the effective viscosity and elastic constant. For high quality electronic grade quartz the value of τ can be taken as 1.6×10^{-14} seconds, equivalent to a Q_1 value at 10 MHz of approximately 10^6. From Eqn (2.55), Q_1 is inversely proportional to frequency and so will be 10^7 at 1 MHz and 10^5 at 100 MHz.

Table 7.3 shows the theoretical minimum values of the motional resistance for the crystals whose C_1 values were tabulated in Table 7.2, calculated from the time constant $\tau = R_1 C_1$ and assuming only the intrinsic material losses to be present. Also given are the maximum ESR values taken from standard commercial crystal specifications, and typical values obtained in a production environment. At the lower frequencies, mounting losses and atmospheric damping are important factors in degrading the attainable Q,

Table 7.3 Motional resistances (ESRs)

Frequency and overtone	R_1 (ideal)	R_1 (max)	R_1 (typical)
5 MHz plano-convex fundamental	2.3	60	20
15 MHz plano-plano fundamental	0.6	20	10
45 MHz third overtone	8	40	20
75 MHz fifth overtone	27	60	40
105 MHz seventh overtone	107	120	—
135 MHz ninth overtone	200	—	—

7.4 FREQUENCY-TEMPERATURE CHARACTERISTICS

Bechmann's power series expansion of the frequency–temperature characteristics of AT-cut resonators has already been discussed in Section 2.6. Bechmann's values for the coefficients of the various terms in the power series were obtained by analysis of the results for a wide range of different resonator frequencies and designs, and in practice have to be treated with caution when dealing with specific cases. Most importantly, Bechmann gives as the reference angle for his results the single value $-35.25°$, whereas in a particular production environment a whole range of reference values has to be used, depending upon the particular design in question.

In most cases it is found that provided the appropriate reference angle is adopted, the normalized curves (Fig. 2.11) resulting from Bechmann's data can still be used to determine the effect of changes in angle from the reference, and also to determine the angular tolerances necessary to meet particular specifications. Figure 7.4 shows curves, derived from those of Fig. 2.11, which show the trade-offs between temperature range, frequency tolerance and angle tolerance. The lower curve shows the best possible theoretical frequency stability for a given temperature range, allowing zero tolerances on the cutting angle, and the upper curves show how the stability requirement must be relaxed in order to allow some tolerance on the orientation. Manufacturing costs rise steeply as the angle tolerance is reduced,

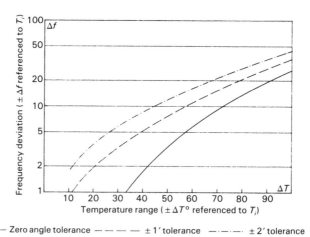

Fig. 7.4 Temperature range vs frequency deviation vs angle tolerance.

making it wise not to overspecify in respect of frequency–temperature tolerance requirements.

The *inflection temperature* is by definition that temperature at which the second derivative of the frequency–temperature curve vanishes. In terms of the power series coefficients in Eqn (2.74), if T_i is the inflection temperature, then T_i is given by

$$T_i - T_0 = -b/3c \qquad (7.5)$$

with T_0 the reference temperature. The inflection temperature is mid-way between the upper and lower turning points of the frequency–temperature curve, and the deviations at the turning points are symmetrical with respect to the inflection point.

A standard method of specifying the allowed variation of frequency with temperature is to state a frequency tolerance of $\pm X$ ppm with respect to the actual frequency measured at a specified reference temperature. If the reference temperature is chosen to coincide with the inflection temperature, then the response of each unit will be symmetrical about the reference, so that if the maximum deviation on the high temperature side is Y ppm, the corresponding deviation on the low temperature side will be $-Y$ ppm. From the point of view of the crystal manufacturer, this is the optimum situation since it maximizes the allowed angle tolerance. This follows because every normalized curve that has a total deviation between the turning points of less than $2X$ ppm will meet the specification, whereas if the reference point is displaced from the inflection point, some of these same curves will fail the specification either at high or low temperatures. This is illustrated in Fig. 7.5, which shows two plots of the same normalized curve; the broken line referred to the inflection temperature as reference, the solid line to a reference temperature below the inflection point. The asymmetry in the second curve is obvious.

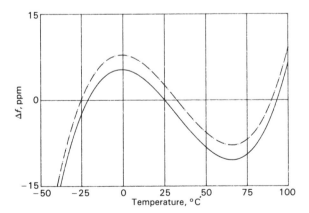

Fig. 7.5 Normalized curves referred to 25 °C and $T_i = 32$ °C.

The actual inflection temperature for AT-cut resonators depends on both overtone and blank geometry. For flat fundamental units, T_i varies within a degree or so of 27°C, depending on the precise angle of cut. For overtone units, T_i is a little higher, around 28°C, but for contoured units is much higher. Plano-convex crystals typically have T_i about 32°C, the precise value depending on the steepness of the contour, whereas bi-convex crystals (or alternatively very steeply contoured plano-convex units) can have T_i around 35°C. In such cases, the asymmetry in the curves when referred to a reference temperature of say 25°C can severely limit the allowable angle tolerance in cutting, and it is then advisable for the user to consider the adoption of a higher reference temperature, or else, what is the same thing in the end, to consider the use of an offset frequency tolerance referred to a lower temperature.

A further point that should be borne in mind in applications is that the reference angle for crystals operated with a load capacitance C_L has to be adjusted depending on the value of C_L. This adjustment results from the shift in the operating point of the crystal between the series resonance and anti-resonance frequencies, which have, as pointed out in Chapter 2, quite different temperature coefficients. The adjustment is only generally significant for flat fundamental mode designs, but in those cases a significant difference in the frequency variation over the temperature range can be expected if a unit designed to operate at series resonance is operated with load capacitor, or vice versa. Figure 7.6 illustrates the case of a 10 MHz fundamental mode unit at series resonance, and with a 30 pF load capacitance.

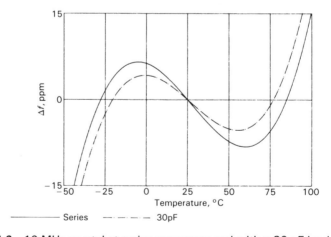

Fig. 7.6 10 MHz crystal at series resonance and with a 30 pF load capacitor.

7.5 UNWANTED RESPONSES

Following the discussion of Chapter 3, unwanted responses in AT resonators can be said to be due either to inharmonic modes found at frequencies just above the desired response and having similar temperature characteristics to the wanted mode, or to flexural, face shear, or extensional modes which can essentially occur at any frequency and have large temperature coefficients. In poorly designed oscillator crystals the inharmonic responses can have motional resistances comparable to or even less than that of the main mode, causing spurious oscillations or frequency jumping in the oscillator circuit. The second group of modes more commonly show themselves in the form of 'activity dips' or 'bandbreaks', that is anomalies in the frequency–temperature characteristics of the crystals as described in Chapter 3.

Activity dips are more likely to be observed in low-frequency designs where the coupling to unwanted modes is stronger, but with proper choice of blank dimensions and contour should not cause a serious problem. The characteristics of inharmonic modes in fundamental mode resonators depend on whether the blank is contoured or not. In the former case, the spacing of the inharmonics from the main mode is controlled by the steepness of the contour, and the location and level of the modes is predictable and repeatable. For flat fundamental designs, the inharmonic modes generally occur considerably closer to the main response but their precise location is difficult to predict. Proper use of the energy trapping criteria of Chapter 3 can result in resonators that are practically free from inharmonic responses, which is necessary in filter crystal applications. For oscillator applications, the energy trapping rules are often over-ridden by requirements for large C_1 values and low R_1 values, but nevertheless reasonable compromises between these requirements and unwanted mode suppression can usually be made.

In the case of overtone crystals, the energy-trapping rules become much more difficult to apply as the overtone order N increases. Consequently it is in general true that the suppression of unwanted responses becomes progressively more difficult to achieve as the frequency and the overtone order increase.

7.6 AGEING AND SHORT-TERM FREQUENCY STABILITY

The frequency stability of crystal resonators is usually considered under two heads, the *long-term* stability or *ageing*, and the *short-term* stability. In the former, the frequency drift of the resonator over periods of weeks, months and years is of interest, whereas in the latter it is the fluctuations in frequency between successive measurements at intervals of the order of seconds or fractions of a second that is important. The general topic of frequency stability and the various statistical methods used to describe it are discussed in detail in the books by Kartaschoff (1978) and Gerber and Ballato (1985).

The physical mechanisms responsible for the short-term instabilities of resonators are much less well understood than those responsible for their long-term drift. As stated in Chapter 9 in reference to the short-term stability of crystal oscillators, the best that can be said at this time is that there appears to be a confirmed correlation between crystal Q and at least one component of the observed instability, but that otherwise the short-term stability seems to depend on such process-dependent factors as surface finish, cleanliness and electrode adhesion.

The long-term stability of crystal resonators has been the topic of extensive research over many years and the chief mechanisms responsible are well understood. These are reviewed in detail by Gerber in Gerber and Ballato (1985), but can be summarized as follows:

(1) stress effects, for example stress relaxation in the mounting structure and the electrode films;
(2) mass loading effects, for example the adsorption of contaminants from the atmosphere;
(3) material effects, such as chemical reactions at the electrode-quartz interface, or structural relaxations in the quartz itself;
(4) other effects, such as changes in hydrostatic pressure due to leaks in the enclosure.

In high-precision resonators, all these factors are important, but in the low to medium precision units that form the vast bulk of resonators used, the first two factors are most critical. The vital importance of cleanliness can be seen from calculating the mass loading effect of a monomolecular layer on the surface of a typical resonator. Gerber (Gerber and and Ballato, 1985) quotes a frequency shift of F ppm in an F MHz resonator as being caused by the adsorption or desorption of contamination equal in mass to $1\frac{1}{2}$ monolayers of quartz, that is 5 ppm in a 5 MHz resonator for example. Even assuming that the blank cleaning process is effective initially, the subsequent sources of contamination in a typical crystal production facility are manifold, including oil molecules from mechanical and diffusion pumps, human skin oils, airborne particles, outgassing from the bonding paste during curing, and outgassing from the crystal holder pieceparts during and after sealing. Although the expense of a complete clean room facility, or of an in-line processing system in which all process steps after cleaning are carried out without breaking vacuum, is probably not justified except for the highest precision requirements, it is nonetheless essential in crystal processing to take all possible practical precautions against contamination. If this is done and proper use is made of high-temperature bakeout procedures before critical process steps, ageing rates of typically 2 or 3 ppm in the first year can be achieved in volume production using resistance weld or cold weld holders.

7.7 NON-LINEAR EFFECTS

The principal non-linear effects observed in crystal resonators can be classified into three groups, those occurring at small signal levels, those occurring at large signal levels, and those involving small signals superimposed on a larger static deformation. In practice, the first group is the more troublesome, since the relevant phenomena seem to be process dependent and relatively difficult to quantify. Comprehensive reviews of the large signal effects and those involving small signals superimposed on large static biases have been given by Gagnepain and Besson (1975) and Gagnepain (1981). These effects involve the non-linear behaviour of quartz as described by the field equations developed in Appendix 3, and include the shift in resonator frequency with signal level (frequency–amplitude effect), the shift in frequency with applied stresses (force–frequency effect) and the shift in resonator frequency with an applied electric field.

The most important small signal effect is the increase in motional resistance that may be observed when the crystal current is reduced to very low levels, commonly known as the 'second level of drive' effect. Since when a crystal oscillator is first switched on the only signals present are noise signals at low level, this increased resistance at low drive can lead to failure of the oscillator to start. In some cases, the phenomenon is only observed after the crystal has been inactive for some time, and can be rectified temporarily by operating the crystal in a high drive level oscillator circuit. The crystal is then colloquially known as a 'sleepy' crystal. The dependence of motional resistance on drive level at low signal levels seems to be entirely dependent on the degree of surface finish of the resonator, its cleanliness and the quality of the electrode film (Bernstein, 1967, Nonaka *et al.*, 1971). It seems probable also that these same factors are responsible for the intermodulation observed in crystal filters at low signal levels (Chapter 10).

8 Measurements

8.1 INTRODUCTION

The quantities to be measured in the determination of the performance of a crystal resonator are the frequency of the resonator under the specified operating conditions, its motional resistance and, optionally, the values of the motional parameters L_1 and C_1 and the shunt capacitance C_0. It may, in addition, be necessary to determine the location and level of unwanted modes. Some or all of these measurements may need to be repeated to determine the variation of selected parameters with such factors as temperature, time and drive level. The range of values of the parameters involved is such that until very recently it was necessary to use measurement techniques and instruments specifically tailored to quartz resonators, but the present day availability of computer controlled test equipment has made it feasible to use standard instruments such as network analysers for crystal measurements.

The methods used for crystal measurements fall into four groups, based, respectively, on crystal impedance meters, bridge techniques, transmission networks, and automated measuring systems. A full review of all these methods is provided by Hafner in Gerber and Ballato (1985), so the discussion here will be limited to a brief outline of the commonly used techniques and of the possibilities offered by the new automated systems.

8.2 CI METERS

The CI meter or *crystal impedance* meter is essentially a crystal-controlled oscillator designed to operate at series resonance, either with or without a series load capacitor. Figure 8.1 shows a simplified block diagram in which the crystal appears in the series arm of a feedback network. The principle of measurement is to find by repeated substitutions and adjustments of the oscillator tuning, a resistor R and a frequency f such that the signal level and frequency are unchanged by substitution of the crystal by the resistor. The series resonance frequency of the crystal and its resistance at that frequency are then precisely f and R.

In practice, the substitution process is very laborious, and is short-circuited by arbitrarily selecting a resistor close to the anticipated crystal resistance and

ZERO PHASE MEASUREMENT SYSTEMS

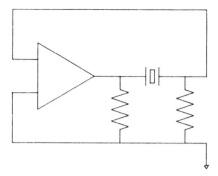

Fig. 8.1 CI meter schematic.

tuning the CI meter to the nominal crystal frequency with this resistor in circuit. The crystal is then substituted for the resistor and the oscillator frequency taken as the crystal frequency. Clearly this method is not so accurate as the full substitution method, although quick and convenient for most practical purposes. Even assuming the full procedure is used, the CI meter still has disadvantages that have resulted in the IEC adopting as a standard a measurement technique based on the use of a transmission network.

8.3 ZERO PHASE MEASUREMENT SYSTEMS

Figure 8.2 shows a block diagram of the basic method adopted by the IEC as an international standard for crystal measurements in the frequency range up to 125 MHz. The method is fully described in IEC Publication 444 (IEC, 1973), the key element in the system being the pi-network, whose schematic is

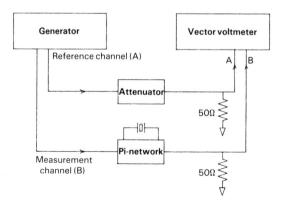

Fig. 8.2 Zero phase measuring system.

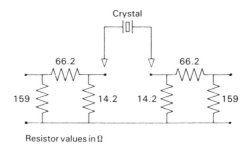

Fig. 8.3 Pi-network schematic.

shown in Fig. 8.3. The pi-network is designed so as to present to the crystal unit precisely defined low impedances of 12.5 ohms, and at the same time to match to standard 50 ohm coaxial cables. When the crystal is replaced by a short-circuit, the pi-network has a loss of 29.6 dB between 50 ohm terminations, and to ensure that the reference and test channels present signals at approximately the same level to the vector voltmeter, an attenuator of approximately this value is placed in the reference channel. As with the CI meter, the zero phase method is in principle a substitution method. A reference resistor is used to set zero phase at the nominal frequency, and is then replaced by the crystal. The generator frequency is then adjusted (manually or often automatically) until the zero phase condition is restored, and the set frequency taken to be the zero phase frequency of the crystal unit. The resistance of the crystal can be calculated from the ratio of the test-channel voltage V_B with the crystal in circuit to the corresponding voltage V_{BS} with the crystal replaced by a short-circuit

$$R = 25\{(V_{BS}/V_B) - 1\} \tag{8.1}$$

The highest accuracy for the frequency measurement is obtained when the reference resistor used has the same value as the crystal resistance, but in any event, the accuracy of the zero phase method depends crucially on the construction of the reference resistors and the pi-network itself.

As in CI meter measurements, the determination of the motional parameters requires several frequency measurements to be made under different conditions, with the parameter values following by calculation. For example, if measurements are made of the load resonance frequencies f_A and f_B corresponding to load capacitors C_A and C_B, the motional capacitance C_1 is given by

$$C_1 = 2(C_A - C_B)(f_B - f_S)(f_A - f_S)/f_s(f_B - f_A) \tag{8.2}$$

where C_0 and f_s are the static capacitance and series resonance frequency respectively. Alternatively, if IEC Publication 444-2 (IEC, 1980) is followed and measurements are made of the frequencies f_1 and f_2 corresponding to phase offsets of ± 45°, then C_1 can be calculated from

$$C_1 = (f_2 - f_1)/(2\pi R_{eff} f_s^2) \qquad (8.3)$$

where the effective resistance R_{eff} is just the sum of the crystal resistance calculated in Eqn (8.1) and the pi-network terminating impedances, that is $R_{eff} = R + 25$.

The zero phase system can in principle measure unwanted modes with no more difficulty than encountered in measuring the principal crystal response, in contrast to CI meters, in which unwanted mode measurement is essentially beyond the capability of the system. The zero phase method has in addition the advantage that since the pi-network is matched to standard coaxial cables, it can be situated remote from the rest of the test system, making it relatively easy to configure environmental test systems. In particular, the pi-network can be located inside an oven to allow precise measurements of frequency–temperature characteristics.

8.4 AUTOMATED MEASUREMENT SYSTEMS

Although the absolute accuracy of the pi-network zero phase method may be questioned insofar as it depends critically on the reference resistors, the method nevertheless provides a workable standard within its operating frequency range. This is, however, limited to 125 MHz at best. Straightforward extensions to higher frequencies are complicated by the difficulty of constructing pi-networks and reference resistors that do not introduce unacceptable phase errors, and more fundamentally by the fact that at higher frequencies and overtones, zero phase frequencies may not exist. As shown in Section 6.3, when the figure of merit M of the crystal is less than the critical value 2, the crystal susceptance does not change sign, that is, there is no frequency interval where the crystal has an inductive impedance and there are no zero phase frequencies. Also, when M is only marginally greater than 2, the zero phase frequency differs appreciably from the resonance frequency of the motional arm of the crystal, which is usually of more significance.

There are several possible approaches to the problem of measurements at these higher frequencies:

(1) physical compensation of the crystal C_0, by one of several alternative means, to allow the continued use of the basic phase zero system;
(2) balanced bridge techniques;
(3) automatic network analyser techniques.

All of these have been proposed as candidates for a standard method of measurement at frequencies beyond the present limit of 125 MHz, but as yet no decision has been made as to which approach, if any, will be implemented. In terms of ease of use as a production tool as opposed to a standard, physical

C_0 compensation has considerable attraction in terms of speed and cost, but for standards purposes, the third option has the advantages of using readily available instruments, of allowing system calibration with traceable impedance standards, and of allowing the use of automated error correction procedures. It would therefore seem probable that at some time in the future a system of this general type will be selected as an international standard for crystal measurements.

Part 4
Crystal oscillators and filters

9 Quartz crystal oscillators

9.1 INTRODUCTION

An electronic oscillator can be characterized as a device for producing ac power at a specific frequency from a dc power source. *Harmonic* oscillators are further characterized by the presence somewhere within the device, not necessarily at the output terminals, of a voltage or current with a nominally sinusoidal waveform. A simple model of a harmonic oscillator consists of a (linear) amplifier and a passive feedback network connected as in Fig. 9.1. When the output from the feedback network is fed back to the input of the amplifier with the correct amplitude and phase, then sustained oscillations can occur. If the amplifier has a voltage gain A and the feedback network has a voltage ratio β, then the necessary condition for sustained oscillation is $A\beta = 1$.

Both A and β will generally be complex functions of the angular frequency ω, so that the complex relation $A\beta = 1$ represents two real conditions. If A, β are written $|A|\exp(j\phi_A)$ and $|\beta|\exp(j\phi_\beta)$ so that ϕ_A and ϕ_β represent the phase shifts through the amplifier and feedback network respectively, then the two conditions for oscillation are

$$\phi_A + \phi_\beta = 2N\pi \tag{9.1}$$

$$|A||\beta| = 1 \tag{9.2}$$

Equation (9.1), with N an arbitrary integer, simply states that the total phase shift around the oscillator must be an integral multiple of 2π or $360°$,

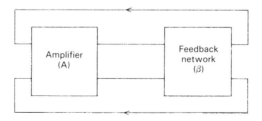

Fig. 9.1 Feedback oscillator schematic.

and Eqn (9.2) requires that the magnitude of the amplifier gain must be sufficient to compensate for the loss in the feedback network.

These conditions are those required for steady-state oscillations. When the amplifier is first turned on, then the only signal present will be white noise, mainly due to the active devices in the circuit. In this initial state, the conditions necessary for the build-up of oscillations at a particular frequency ω are that the phase condition Eqn (9.1) be satisfied, and that the loop gain $|A||\beta|$ be *greater* than 1. In order for the amplitude of the oscillations not to increase without limit, it is then necessary to assume that the voltage gain A of the amplifier decreases with amplitude, so that at some finite signal level the gain condition Eqn (9.2) is satisfied.

These basic considerations are common to all types of harmonic oscillator, including crystal oscillators, and there are excellent texts available dealing with the elaborations necessary in specific cases. In particular, the recent books by Frerking (1978), Matthys (1983) and Parzen (1983) can be recommended for the details of crystal oscillator design. The review articles by Smith and Frerking in *Precision Frequency Control* (Gerber and Ballato, 1985) should also be consulted for detailed references, particularly in the context of applications requiring the highest stability. For present purposes, it is sufficient to consider an elementary small-signal model of a single transistor oscillator to illustrate the commonly used circuit configurations and their differences and similarities.

9.2 ELEMENTARY CIRCUITS

Figure 9.2 shows in schematic form a basic single transistor oscillator, with all power supply and biasing circuits omitted for clarity. Replacing the transistor by its small-signal hybrid-pi equivalent circuit leads to Fig. 9.3, and absorbing the passive equivalent circuit elements into the impedances Z_1, Z_2

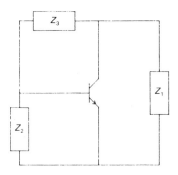

Fig. 9.2 Single transistor oscillator.

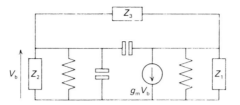

Fig. 9.3 Single transistor oscillator equivalent circuit.

and Z_3 finally leads to the circuit of Fig. 9.4. If I_1 and I_2 are the currents through Z_1 and Z_2, respectively, then by inspection

$$I_1 Z_1 - I_2(Z_2 + Z_3) = 0 \tag{9.3}$$

$$-g_m V_b = I_1 + I_2 \tag{9.4}$$

$$V_b = I_2 Z_2 \tag{9.5}$$

where g_m is the transconductance of the transistor and V_b the base-emitter signal voltage.

Substituting Eqn (9.5) into Eqn (9.4) leads to

$$I_1 + I_2(1 + g_m Z_2) = 0 \tag{9.6}$$

Equations (9.3) and (9.6) constitute a pair of simultaneous equations for the currents I_1 and I_2, and the condition for non-trivial solutions to exist is that the determinant of the coefficients vanishes, that is, that

$$Z + g_m Z_1 Z_2 = 0 \tag{9.7}$$

where Z has been written for $Z_1 + Z_2 + Z_3$.

Equation (9.7) takes on a particularly simple form when the reactive parts of Z_1 and Z_2 are much larger than their resistive parts, that is when Z_1, Z_2 can be replaced by jX_1 and jX_2. Then writing Z_3 as $R_3 + jX_3$ and separating the real and imaginary parts of Eqn (9.7) gives

$$X_1 + X_2 + X_3 = 0 \tag{9.8}$$

$$g_m = R_3 / X_1 X_2 \tag{9.9}$$

Equation (9.8) determines the frequency of oscillation as the series resonance

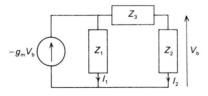

Fig. 9.4 Simplified equivalent circuit.

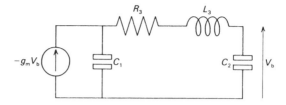

Fig. 9.5 Pierce type circuit.

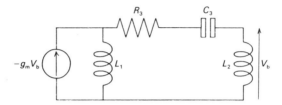

Fig. 9.6 Miller type circuit.

frequency of the loop Z_1, Z_2 and Z_3, and Eqn (9.9) determines the transconductance g_m required to maintain the oscillation. As expected, the larger the value of R_3, the larger is the required value of g_m.

It further follows from Eqn (9.9) that since g_m and R_3 are necessarily positive, so too is the product $X_1 X_2$. Therefore X_1 and X_2 must both have the same sign, and it then follows from Eqn (9.8) that X_3 must have the opposite sign. Consequently, the circuit of Fig. 9.4 has the two possible realizations shown in Figs. 9.5 and 9.6, where it is to be understood that the L, C and R elements shown are to be interpreted as the effective values of what in real oscillators would be combinations of circuit elements. Both realizations have been utilized in crystal oscillators, with that of Fig. 9.5 leading to the commonly used *Pierce*, *Colpitts* and *Clapp* circuits, whereas that of Fig. 9.6 leads to the now rarely used *Miller* circuit, which will not be further considered. The Pierce, Colpitts and Clapp circuits, although sharing the same small-signal ac equivalent circuit, differ in the location of the ac ground point. Consequently, their actual physical realizations and the details of their design and performance differ markedly. In the Pierce circuit, the emitter is at ac ground, while in the Colpitts and Clapp the ground point is at the collector and base respectively. Figures 9.7, 9.8 and 9.9 show the elementary schematics for the three circuits.

The Pierce circuit is generally preferred to the Colpitts on the grounds of ease of design and superior frequency stability (Frerking, 1978). However, the fact that in the Colpitts one side of the crystal is grounded makes this circuit very popular where a number of crystals are to be switched in and out of the circuit.

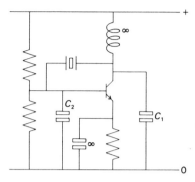

Fig. 9.7 Pierce oscillator.

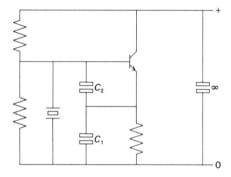

Fig. 9.8 Colpitts oscillator.

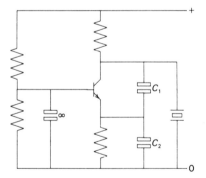

Fig. 9.9 Clapp oscillator.

9.3 FREQUENCY STABILITY: CIRCUIT CONSIDERATIONS

The primary characteristic of crystal oscillators as compared to other types of electronic oscillator is their superior frequency stability. This results in the first instance from the characteristics of the crystal itself, but nevertheless the sustaining circuit must be carefully designed in order not to degrade the performance of the crystal. This section outlines some of the relevant considerations in the choice of the circuit elements, and the following section covers the key characteristics of the crystal itself.

If the transistor transconductance g_m is assumed to be real and the impedances Z, Z_1, Z_2 and Z_3 are written $R + jX$, $R_1 + jX_1$, etc., then the fundamental condition for oscillation Eqn (9.7) can be separated into its real and imaginary parts

$$\left. \begin{array}{l} R + g_m(R_1R_2 - X_1X_2) = 0 \\ X + g_m(X_1R_2 + X_2R_1) = 0 \end{array} \right\} \quad (9.10)$$

Eliminating g_m leads to the relation

$$X/R = (X_1R_2 + X_2R_1)/(R_1R_2 - X_1X_2) \quad (9.11)$$

Equation (9.11) can be regarded as the determinant of the frequency of oscillation, through the frequency dependence of the various reactances and resistances appearing in the equation. For crystal oscillators of the Pierce family, the crystal impedance is contained in the terms X and R on the left side. Consequently, *in the neighbourhood of the crystal's resonance frequency*, the left side of Eqn (9.11) is strongly dependent on frequency, whereas the right side is to a good approximation independent of frequency. Writing $F(\omega)$ for X/R and F_0 for the right side of Eqn (9.11), then leads to the frequency equation

$$F(\omega) = F_0 \quad (9.12)$$

In the Pierce case, the reactances X_1 and X_2 are both negative, so that assuming R_1 and R_2 to be comparatively small the quantity F_0 is small and positive. By definition, F is the ratio of X to R, and is therefore the tangent of the phase angle ϕ of the loop impedance Z, which is in turn made up of the crystal impedance Z_3 in series with the capacitive impedances Z_1 and Z_2. Consequently Z consists of the crystal in series with an effective load capacitance of reactance $X_1 + X_2$, damped by the resistive components R_1 and R_2. Equation (9.12) then demands that the circuit oscillate at a frequency such that the phase ϕ has a small positive value.

Supposing now that some change in the circuit elements of the oscillator (other than the crystal itself) causes a small variation in the equivalent impedances Z_1 and Z_2 and consequently a small change dF_0 in F_0, then it is evident from Eqn (9.12) that for oscillations to be maintained there must be a corresponding change dF in $F(\omega)$. If the accompanying change in frequency is $d\omega$, then $dF = (dF/d\omega)d\omega$ and

$$d\omega = dF_0/(dF/d\omega) \quad (9.13)$$

Clearly then, to minimize the frequency changes due to incidental changes in

FREQUENCY STABILITY: CIRCUIT CONSIDERATIONS

the oscillator circuit, the quantity $(dF/d\omega)$ must be maximized.

As shown in Section 6.4, a crystal with a series load capacitor is equivalent to a crystal with no load capacitor but with modified parameter values. Also, for a crystal with a good figure of merit M (Section 6.3.1), a resistance in series can be regarded as effectively adding to the motional arm resistance of the crystal. The loop impedance Z is then just equivalent to that of a crystal with parameters given in terms of the actual crystal impedance Z_3 and the impedance Z_1 and Z_2 by

$$C'_0 = C_0 C_L/(C_0 + C_L)$$
$$C'_m = C_m/K^2$$
$$R'_m = K^2 R_m + R_1 + R_2 \quad (9.14)$$
$$Q' = Q K^2 R_m/(K^2 R_m + R_1 + R_2)$$

where C_L is the effective load capacitance; C_0, C_m and R_m are, respectively, the shunt and motional capacitances and the motional resistance of the crystal; and the primed parameters are the effective values in the circuit. The multiplier K is given by

$$K = 1 + C_0/C_L$$

and C_L is

$$C_L = -1/\{\omega(X_1 + X_2)\}$$

with ω the nominal oscillation frequency.

As shown in Section 6.3.3, the phase angle ϕ of the admittance of a crystal unit is given in terms of the figure of merit M and the normalized frequency variable x by

$$\tan(\phi) = (1 + x^2 M^2 - xM^2)/M$$

and its rate of change with frequency by

$$d(\tan(\phi))/d\omega = 2(Q/\omega_s)(2x - 1)$$

In the present case, the phase angle of the loop *impedance* Z is therefore given by

$$F = \tan(\phi) = -(1 + x^2 M^2 - xM^2)/M \quad (9.15)$$

and the rate of change with frequency by

$$dF/d\omega = -2(Q/\omega_s)(2x - 1) \quad (9.16)$$

where in Eqns (9.15) and (9.16) the parameters M, Q and x are all appropriate to the effective crystal parameters defined in Eqn (9.14). In particular, the Q factor appearing in Eqn (9.16) is the effective Q defined in the fourth of Eqns (9.14).

Now the condition for oscillation, Eqn (9.12), implies that F be small and positive. From the discussion in Section 6.3.2, it then follows that the normalized frequency x must be in the vicinity of one of the zero phase frequencies $x = M^{-2}$ or $x = 1 - M^{-2}$. Since in the latter case the effective crystal resistance is much higher than in the former (Section 6.3.7), it follows that x must in fact have a value slightly above M^{-2}, so that F is approximately $(xM^2 - 1)/M$. Then to maximize the magnitude of $dF/d\omega$ in Eqn (9.16), it is clear that on the one hand the effective Q must be as large as possible, but on the other hand x must be kept as close as possible to the limiting value of M^{-2}. The latter condition is equivalent to the condition that the value of F and hence of F_0 be minimized.

For a given crystal Q, it follows from the fourth of Eqns (9.14) that to prevent the effective Q from being significantly degraded by the circuit, the resistive components R_1 and R_2 of the impedances Z_1 and Z_2 must be kept very much less than the effective motional resistance of the crystal K^2R_m. Assuming that R_1 and R_2 have been reduced to a minimum, the second condition for maximizing the frequency stability implies that the reactive components X_1 and X_2 have to be chosen so that F_0 is minimized. Noting that the total reactance $X_1 + X_2$ is effectively fixed by the need to operate the crystal on a specified load reactance, the minimization of F_0 has to be subject to this constraint, so that any change dX_1 in X_1 has to be accompanied by an equal and opposite change $dX_2 = -dX_1$ in X_2. For small R_1 and R_2, F_0 reduces to

$$F_0 = -R_2/X_2 - R_1/X_1$$

so that

$$\begin{aligned} dF_0/dX_1 &= (R_2/X_2^2)(dX_2/dX_1) + R_1/X_1^2 \\ &= R_1/X_1^2 - R_2/X_2^2 \end{aligned} \quad (9.17)$$

The extreme value of F_0 occurs when dF_0/dX_1 vanishes, that is when

$$(X_1/X_2)^2 = R_1/R_2 \quad (9.18)$$

If as is frequently more convenient the impedance Z_1, Z_2 are expressed in terms of the equivalent admittances $G_1 + jB_1$ and $G_2 + jB_2$, then this expression can be rewritten as

$$(B_1/B_2)^2 = G_1/G_2 \quad (9.19)$$

Either of Eqns (9.18) or (9.19) then allow the optimum choice to be made for the ratio of the reactive components of Z_1 and Z_2.

9.4 FREQUENCY STABILITY: CRYSTAL CHARACTERISTICS

As shown in Section 9.3, an appropriate choice of the circuit elements in a crystal oscillator can minimize the effect of slight changes in the elements on the output frequency of the unit. On the other hand, slight changes in the crystal frequency itself will be reflected directly in changes in the output frequency. Such instabilities can be classified under three main headings: the *short-term* frequency stability, the *long-term* frequency stability or *ageing*, and the *temperature* stability.

Short-term stability refers to the variation in output frequency of the oscillator observed in successive measurements. The usual measure of this quantity is the standard deviation of a number of frequency readings, each reading being effectively the average frequency over a period of time known as the *sampling interval* τ. The standard deviation is then found to depend on the length of the sampling interval, as shown schematically in Fig. 9.10. The different regions of the typical curve shown can be associated with different statistical noise processes, but the further association of these processes with actual physical mechanisms in the resonator or the sustaining circuit is still a subject of active research (Gerber and Ballato, 1985). There appears to be a confirmed correlation between resonator Q and that component of the frequency instability usually described as '1/f' or 'flicker' noise, but other than this only the general statement that short-term stability appears to depend on such process-dependent factors as surface finish, cleanliness, and electrode adhesion can be made. Since these factors also influence such non-linear phenomena as the dependence of crystal frequency and resistance upon drive level, it is sometimes possible to correlate these effects with observed short-term frequency stabilities and use them to select crystals that will have improved short-term behaviour. It is also empirically established that where short-term stability is of importance, the crystal unit should be operated at a relatively high drive level. Unfortunately, this conflicts with the need to operate at very low levels of drive to achieve the best possible long-term ageing behaviour, discussed next.

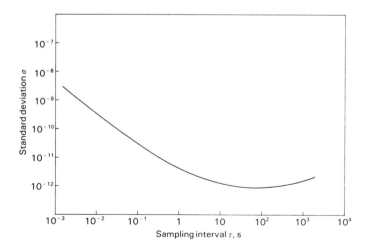

Fig. 9.10 Short term frequency stability.

The long-term stability of oscillators or crystal units refers to their frequency drift over periods of months or years, and is usually expressed as ppm per year or parts in 10^{-8} per day. In a properly designed oscillator, where the effects of component variations have been minimized, the long-term stability is almost entirely that of the crystal unit, which has been discussed in Chapter 7.

As in the case of the long-term stability, in a properly designed oscillator the frequency–temperature stability will be that of the component crystal. Exceptions to this rule would occur if, for example, inductors were used to extend the frequency adjustment range of the oscillator without due attention to their temperature coefficients, or if the crystal were operated at a load capacitance much different from the specified nominal value. The frequency–temperature characteristics of the AT-cut resonator family are discussed in Chapter 7, and the techniques of temperature compensation and control, used to obtain oscillators with superior temperature stabilities to those that can be obtained from the crystal alone, are briefly discussed in the following section.

9.5 TYPES OF CRYSTAL OSCILLATOR

Crystal oscillators are available for a wide range of applications, with stabilities ranging from 0.1% for simple timing functions, as in microprocessor clocks, to 5×10^{-11} for secondary frequency standards or reference frequency sources for satellite navigation systems. Prices correspondingly range from a few pounds to many thousands of pounds for the most sophisticated devices.

Despite the great variety of available oscillator types, it is useful to categorize crystal oscillators into four main groups. These are *simple packaged oscillators, temperature compensated oscillators, oven controlled oscillators*, and *voltage controlled oscillators*. These groupings overlap, for instance in such cases as voltage controlled temperature compensated crystal oscillators, and take no account of the differences that may exist between oscillators in the same group in respect of such factors as output waveform and level, power supply requirements, provision of auxiliary outputs at multiples or sub-multiples of the nominal frequency, and so on. Nevertheless the classification is useful for purposes of description, and follows that of the IEC Publication 679-2, *Quartz crystal controlled oscillators, Part 2: Guide to the use of quartz crystal controlled oscillators* (IEC, 1981).

9.5.1 Simple packaged crystal oscillators (SPXOs)

SPXOs are crystal oscillators in which no provision is made either for temperature compensation or temperature control, so that the frequency–temperature stability is essentially that of the crystal itself. The AT-cut crystal is generally adopted for use in such oscillators, with frequency division being used to obtain output frequencies below the practical AT manufacturing range. As discussed in Chapter 7, the characteristics of the AT-cut are such that good frequency stability can be achieved over wide temperature ranges, with ± 10 ppm over a temperature range of −20 to +70°C being typical. An exception to this general rule is the recent announcement (Ochiai *et al.*, 1986) of packaged oscillators employing the miniature GT cut crystal described in Section 1.4.6. These units are claimed to have stabilities of ± 2.5 ppm over a temperature range of −30 to +70°C, but are so far restricted to a frequency range of 1 to 3 MHz.

By far the most common application of SPXOs is in providing timing signals in digital electronic equipment. Such oscillators are termed *crystal clock oscillators* and are produced in high volume in standard dual-in-line packages, with outputs tailored to be directly compatible with TTL or CMOS logic circuits. Clock oscillators are produced in the freuency range up to 70 MHz, with some recent devices extending the upper limit to 100 MHz by using fifth overtone crystal units (Oita, 1986). The manufacture of clock oscillators makes full use of hybrid circuit techniques. The active components in the maintaining circuit are integrated on a single IC chip and the crystal blank is mounted on pillars on the substrate to allow direct adjustment of the crystal (and hence the oscillator) frequency. The need for a separate package for the crystal is eliminated by using resistance-welding techniques to provide a hermetically sealed enclosure for the complete oscillator assembly. Typical stabilities for this type of clock oscillator are ± 50 or ± 100 ppm over a temperature range of 0 to 70°C.

9.5.2 Temperature compensated crystal oscillators (TCXOs)

The AT-cut quartz resonator provides excellent frequency stability with respect to temperature variations as compared to most alternatives. For example, over an operating temperature range of −10° to +60°C, a stability of ± 5 ppm can be routinely achieved. With additional care and some penalty in cost, the temperature range can be marginally extended or, alternatively, the stability marginally improved. To obtain further significant improvements, however, necessitates a different approach. For example, to obtain a stability of ± 5 ppm over the temperature range −30° to +60°C is not practically possible with an unaided AT-cut crystal, even though it is theoretically feasible. There are two main approaches to the problem of improving upon the stabilities available from an unaided crystal. In the first, to be considered in the following section, the crystal and possibly the oscillator circuit itself are placed in a temperature-controlled environment provided by some sort of oven. In the second, temperature sensitive elements are introduced into the oscillator circuit in such a way that the variation in crystal frequency with temperature is counterbalanced by variations in the effective load reactance of the crystal.

The load reactance is usually a capacitor with a value in the region of 20 pF. 'Active' compensation techniques utilize a varactor diode to provide the load capacitor. Variation of the bias voltage applied to the varactor results in a variation of load capacitance and hence a corresponding variation in the crystal frequency. 'Passive' compensation techniques realize the crystal load capacitance as a network of capacitors, resistors and thermistors whose reactive component displays the desired temperature behaviour.

The essence of the active method of compensation is to tailor the varactor bias circuit so that the bias voltage produced as a function of temperature is precisely that required to produce frequency changes in the crystal equal and opposite to those induced by the temperature variations themselves. Active compensation techniques can be further classified as analog or digital, depending on the method adopted to generate the required bias voltage. In the classical, analog, approach the bias circuit is made up of a network of resistors and temperature-dependent elements. The latter are most commonly thermistors, but other approaches have been tried. In the digital approach, the bias voltages required at several discrete points across the temperature range are stored in a PROM. The signal from a suitable temperature-sensing device is then converted to digital form and used to address the PROM, with the bias voltage then being applied to the varactor diode as in the analog case.

It is generally the case that the active techniques provide better performance, but at some cost in terms of additional power consumed in the bias circuit and in additional components. Using analog techniques, about the best that can be achieved is a stability of 0.5 ppm over a temperature range of -30 to $+80°C$, with the limiting factors being the ageing of the crystal and also the presence of perturbations in the frequency–temperature characteristic of the crystal. As usual, there is a 'trade-off' between temperature range and achievable stability. With digital techniques, there is in principle no limit to the degree of compensation possible aside from the ageing of the crystal and the repeatability of its frequency–temperature characteristic, but in practice there are limitations owing to the finite precision of the digital devices used. Stabilities of better than 0.1 ppm over wide temperature ranges have been reported (Frerking, 1978). However, it is presently the case that digitally compensated TCXOs remain very much more expensive than their analog counterparts, and are much less readily available.

9.5.3 Oven controlled crystal oscillators (OCXOs)

TXCOs provide an excellent solution to the problem of providing a medium to high precision frequency source in a wide range of applications, particularly those where constraints on size and power consumption effectively rule out the use of oven controlled oscillators. A typical application would be as a reference frequency source in a synthesized portable or mobile radiotelephone. Nevertheless, for the highest precision requirements, the use of an oven to provide a stable operating temperature remains mandatory, despite the disadvantages mentioned of increased size and power consumption.

The ovens used in OCXOs can generally be classified as either 'on/off' or 'proportionally controlled' types. In the former, a sensor, such as a bimetallic strip or a mercury-contact thermometer, is used to detect changes in the oven temperature and switch the oven heater on or off as required. This has the advantage of simplicity, but on the other hand leads to relatively large variations of the oven temperature between the on and off cycles, and additionally the set point temperature is liable to drift over long periods of time. Consequently for precision oscillators, the second type of oven control is preferred, in which the oven temperature is continuously monitored and the heater current adjusted according to the difference between the actual oven temperature and the nominal set point temperature. This is usually achieved by using a temperature-sensitive resistance in a bridge circuit and using the output from the bridge to control the heater current.

For the very highest precision, elaborate double ovens are used, with the crystal and essential parts of the oscillator circuit contained within the inner oven, and the oven control and power supply circuits contained in the outer oven. This allows at least an order of magnitude improvement in the temperature stability of the inner oven to be achieved as compared to a single oven. Typical values are 0.5 to 0.01 °C stability for a single oven, and 0.001 °C for a double oven, when operated over wide ambient temperature ranges (cf. Frerking in Gerber and Ballato, 1985).

To take full advantage of these temperature stabilities, the oven operating temperature has to be carefully matched to the turnover temperature of the crystal employed. As already discussed in Section 2.6, it is in this area that the newer doubly rotated crystal cuts have been found to have significant advantages over the AT-cut. Firstly, the slope of the frequency-temperature characteristic in the neighbourhood of the turnover point is smaller, making the error resulting from a mismatch of oven temperature and turnover point less critical, but more importantly the transient frequency changes resulting from the temperature cycling of the oven are also much reduced in the doubly rotated cuts.

9.5.4 Voltage controlled crystal oscillators (VCXOs)

VCXOs are oscillators whose output frequency is capable of being controlled by an external voltage. Typically, such oscillators would be used in phase-locked loop applications, or to allow frequency modulation of the output signal. The frequency variation is usually obtained by varying the bias voltage on a varactor diode in series with the crystal, just as in a TCXO, but with the differences that the control voltage has to be supplied by the user, and usually a rather larger range of adjustment is required than in a TCXO.

This second factor, together with the need to achieve a linear relationship betweeen control voltage and output frequency, implies that often more sophisticated networks than a simple connection of the varactor in series with the crystal are required. The simple formula for the shift in crystal frequency with load capacitance (Eqn 6.2) shows the inherent non-linearity involved with large changes in load capacitance value. Fortunately the non-linearity of the voltage–capacitance characteristic of typical varactor diodes tends to compensate for the non-linearity of the crystal 'pulling' characteristic, so that the overall linearity of the voltage–frequency characteristic is considerably better than that of either the crystal or the varactor considered separately. These points, together with the advantages and disadvantages of using inductors to increase the range of frequency adjustment are discussed in detail by Neubig (1979). In any case, however, it should be noted that in increasing the adjustment range of the output frequency, whether by introducing inductors or by increasing the 'pullability' of the crystal itself, the inherent stability of the oscillator itself is correspondingly decreased.

10 Quartz crystal filters

10.1 INTRODUCTION

Electronic filters are signal-processing devices whose function is to discriminate in some way between the different frequency components of an input signal. *Lowpass* and *highpass* filters are, respectively, filters that transmit signals at frequencies below or above a defined cut-off frequency and attenuate those at frequencies above or below the cut-off. *Bandpass* filters transmit all frequencies between defined upper and lower limits, and attenuate frequencies outside those limits. *Bandstop* filters attenuate frequencies between upper and lower limits and transmit all other frequencies. *Allpass* filters transmit all frequencies, but introduce time delays depending on the frequency of the individual signal components.

Ideally, filters would have zero attenuation in their *passband(s)*, that is those frequency intervals in which signals are to be transmitted, and infinite attenuation in their *stopbands*. Additionally, the phase shift produced by the filter for signals in the passband would be proportional to the frequency so that the *group delay*, which is the derivative of the phase with respect to the frequency, would be constant in the passband. This would then allow the undistorted transmission of a signal made up of components with frequencies lying entirely within the passband. In practice, these characteristics of an ideal filter are not realizable with a finite number of components, so that real filters have to be defined in terms of the following parameters (or some equivalent set):

(1) maximum attenuation allowed in the passband;
(2) minimum attenuation required in the stopband;
(3) width of the *transition region* from those points in the passband where the maximum attenuation allowed is last achieved, to those adjacent points in the stopband where the minimum attenuation required is first achieved;
(4) allowed deviation from a linear phase or constant group delay characteristic in the passband.

Figures 10.1 and 10.2 illustrate these parameters for the simple case of a lowpass filter.

Generally speaking, the complexity, that is the number of elements, needed in a filter increases with the level of attenuation required in the stopband for a

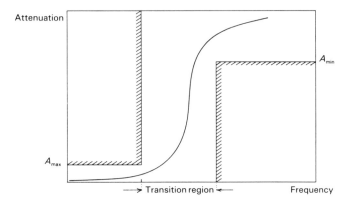

Fig. 10.1 Amplitude specification, LP filter.

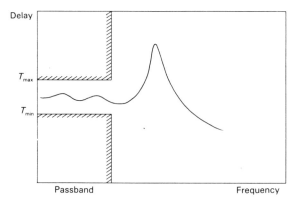

Fig. 10.2 Group delay specification, LP filter.

given transition region, or alternatively, for a given level of stopband attenuation, the complexity increases as the width of the transition region is reduced. Additionally, for the common class of filters known as *minimum-phase* filters, the phase or group delay response is not independent of the amplitude response. Specifically, for such filters, a narrow transition region between pass and stopbands is accompanied by large variations in the group delay near the passband edges, whereas, on the other hand, a flat group delay response implies a rounded passband and a wide transition region. To overcome this requires either additional complexity in the filter, such as the use of allpass sections to *equalize* the group delay response, or else the use of non-minimum phase filters, which implies greater complexity in the filter design process.

Electronic filters can be realized in many different forms, ranging from the

INTRODUCTION

classical *LC filter*, utilizing inductors and capacitors, through *mechanical*, *ceramic* and *crystal* filters, various forms of *microwave* filters, *surface acoustic wave* (SAW) devices, up to active filters, again realized in various different forms, such as *active RC* filters or *switched-capacitor* devices. *Crystal filters* are passive electro-mechanical devices in which crystal resonators are used in place of *LC* resonant circuits. Crystal filters can themselves be divided into two categories, *discrete* and *monolithic* (or *polylithic*). The former use individual crystal resonators as circuit components, whereas the latter use resonators that are acoustically coupled by virtue of their fabrication in close proximity on the same crystal blank. Although resonator materials other than quartz have been used in the past, and are still actively being investigated, quartz remains the most commonly used, thanks to its unique combination of very low losses and high stability with respect to both time and temperature.

The specific applications in which quartz crystal filters come into their own are those requiring narrowband highly selective devices. There have historically been two major application areas, the first being the use of crystal filters in telephone *frequency-division multiplex* (FDM) systems. In such systems, many telephone channels are transmitted simultaneously over a wideband transmission line, such as a coaxial cable, and separated at the receiving end by banks of *channel filters*, each bank consisting of several filters with a nominal bandwidth of approximately 3 kHz and centre frequencies equally spaced across the frequency range occupied by the group of channels. In order to prevent interference between the channels, the filters must be extremely selective, and the high Q of quartz crystal resonators that allowed the construction of such filters was a key factor in the successful implementation of FDM telephone systems. The application of crystal filters in this field was well established by the 1940s and accounted at that time for the vast bulk of crystal filters produced. However, both the introduction of *time-division multiplex* (TDM) systems and the use of *mechanical* rather than crystal channel filters has meant that the importance of the latter has decreased.

The second major application area for crystal filters, and currently most important, is in radio communication and electronic navigation systems. In such systems the prime use is as *intermediate frequency* (IF) filters, although there are some secondary applications such as 'front-end' or *antenna* filters, and occasional applications for bandstop filters. Sometimes also crystals are 'imbedded' in wider band networks to provide sharp attenuation peaks, but neverthless the overwhelming majority of crystal filters are used in narrowband, bandpass roles. In VHF and UHF radio systems, they are used to provide the necessary discrimination between closely spaced communications channels. In HF systems, besides this function, crystal filters are also used to separate the sidebands in SSB transmissions and to act as 'roofing' filters in double conversion systems.

10.2 ELEMENTARY CIRCUITS: DISCRETE CRYSTAL FILTERS

The general problem of the design of a crystal filter falls into two distinct parts. On the one hand, there is the problem of designing a filter network capable of accommodating the particular equivalent circuit of a crystal resonator, and on the other, the problem of designing resonators with the appropriate values for the elements of the equivalent circuit as determined by the network design. Usually, the network design problem is handled with the tools developed for the synthesis of *LC* filters, which in itself is a highly developed subject. The original approach to the design of *LC* filter circuits, the *image parameter* method, has in modern practice been superseded by 'exact' methods which in their full generality rely heavily upon the use of digital computers to carry out the numerical calculations involved.

[Note: there is a very large literature on the subject of network synthesis, although relatively little has been written on the particular subject of crystal filters. For references to the original work in the field and to modern developments, the books by Zverev (1967), Humpherys (1970), Weinberg (1962) and Temes and Mitra (1973) should be consulted. Mention should also be made of the collection of reprints edited by Sheahan and Johnson (1977), which contains several important papers on crystal filters, and of the extensive bibliography on crystal, mechanical and SAW filters contained in Gerber and Ballato (1985).]

The 'exact' methods of filter synthesis can be shown to yield 'optimum' designs, with 'optimum' being defined in terms of specific performance characteristics. However, they do not in general give the designer the same degree of control over the final structure of the filter as would have been implicit in the old image parameter method. Thus it is often the case that in designing a crystal filter, the network provided by an exact synthesis technique only provides a starting point, with more or less extensive network transformations being required to produce a form in which the equivalent circuit of the crystal can readily be accommodated. However, once the network is obtained, its analysis can be straightforward, and the operation of the standard crystal filter configurations can be understood on basic principles, even if the specific choice of element values has to be based on complex calculations.

Considering first discrete crystal filters, that is, filters using individual crystal resonators to replace *LC* resonant circuits, a broad distinction can be made between narrow, intermediate and wideband filters. (In the overall context of electronic filters, all these categories would be treated as narrowband: the present distinction is made strictly within the context of crystal filters.) *Narrowband* units are those that can be realized in principle by crystals and capacitors only; *intermediate* band filters use inductors either wholly or partially to compensate for the crystal static capacitances and other stray capacitances; and *wideband* units use inductors as integral elements of the filter.

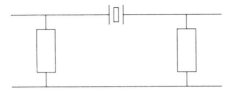

Fig. 10.3 Crystal in series arm of ladder circuit.

Narrowband crystal filters can be realized as either ladder or lattice structures. When a crystal is used in a series arm of a ladder structure as in Fig. 10.3, it is clear that the series resonance frequency of the crystal must lie in or close to the passband of the filter, whereas the anti-resonance frequency must lie in the stopband. For a ladder with a crystal in a shunt arm (Fig. 10.4), the reverse is true, and in either case it follows that the bandwidth of the filter, that is the width of the passband, must be substantially less than the separation of the series and anti-resonance frequencies of the crystal. From Chapter 6, if the motional capacitance and static capacitance of the crystal are C_1 and C_0, respectively, this spacing is given by

$$\Delta f/f_s = C_1/2C_0 \qquad (10.1)$$

For a fundamental mode AT-cut crystal, the capacitance ratio C_0/C_1 is approximately 200, so that from Eqn (10.1) the fractional bandwidth of a narrowband filter using AT-cut fundamentals must be substantially less than 1/400 or in percentage terms substantially less than 0.25%. For overtone crystals, the capacitance ratio reduces with the square of the overtone order, so that with third overtone crystals for example, the maximum narrowband ladder bandwidth must be less than 0.025%.

This bandwidth restriction is a severe limitation on the ladder structure realization. In addition, simple ladders with crystals either in the series arms only, or in the shunt arms only, necessarily have attenuation peaks either above or below the passband which although possibly of use in SSB applications are generally an embarrassment. Both disadvantages are avoided in the lattice arrangement of Fig. 10.5. Although, following convention, only two lattice arms are shown in the figure, the full lattice arrangement is in fact fully symmetrical and balanced, being equivalent to a bridge network with the filter input being applied to one pair of nodes and the output taken

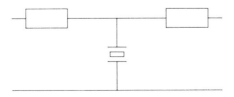

Fig. 10.4 Crystal in shunt arm of ladder circuit.

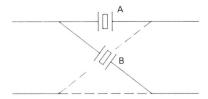

Fig. 10.5 Crystal lattice.

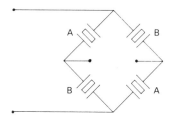

Fig. 10.6 Lattice redrawn as a bridge.

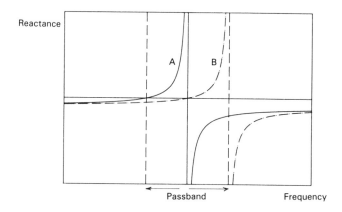

Fig. 10.7 Reactance plot for lattice filter.

from the other pair, as in Fig. 10.6. Clearly, when the lattice arms A and B have equal impedances, the bridge is balanced and there will be no output, that is there will be a pole of attenuation.

An image parameter analysis shows that if the lattice arms A and B have impedances $Z_A = jX_A$ and $Z_B = jX_B$, and if the lattice is terminated in the image impedance $(Z_A Z_B)^{1/2}$, then the attenuation of the lattice is zero when X_A and X_B have opposite signs. If each lattice arm contains a single crystal and the crystal resistances are neglected, then the arm reactances X_A and X_B will be as drawn in Fig. 10.7, where the series frequency of crystal B has been chosen to coincide with the anti-resonance frequency of crystal A. Clearly X_A and X_B have opposite signs from the series frequency of A to the pole frequency of B, so that the bandwidth of the lattice extends to twice the pole-zero spacing of the crystals, and therefore more than twice the bandwidth of a corresponding ladder filter.

In the stopband region, the attenuation of the section depends on the ratio of X_A and X_B, going to infinity when the ratio is unity. In the far stopband, the crystal impedances essentially reduce to the impedances of their static capacitances, so that if these are made the same, then the attenuation in the stopband will tend monotonically to infinity, or in practice, to some large finite value determined by circuit strays and component tolerances. If on the other hand the static capacitances are chosen to have some ratio other than unity, attenuation peaks appear on either side of the passband, their precise location depending on the ratio of the static capacitances. Figure 10.8 shows typical examples of the image attenuation due to a single lattice section.

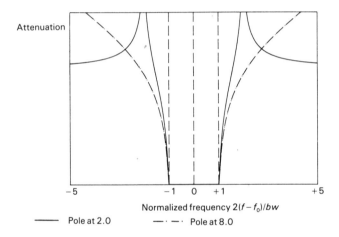

Fig. 10.8 Image attenuation of a lattice section.

Fig. 10.9 'Half-lattice' transformation.

The full lattice of Fig. 10.5 contains four crystals in two identical pairs. Apart from the manufacturing difficulties in producing matched pairs of crystals, it would clearly be a great advantage if the need for identical crystals could be eliminated. This can in fact be done in several different ways, but the most commonly used is the 'half-lattice' transformation of Fig. 10.9,

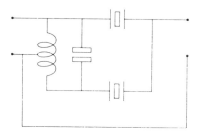

Fig. 10.10 'Half-lattice' with tuned transformer.

whereby the number of crystals is halved at the expense of the introduction of an ideal centre-tapped transformer. In practice, the transformer is realized as a tightly coupled tuned transformer, tuned to the centre frequency of the filter, as in Fig. 10.10. Because of the very narrow bandwidths involved, the effect of the tuned circuit on the filter response can be neglected, except insofar as the finite Q of the inductor introduces losses that add to the losses due to the motional resistances of the crystals.

The flexibility of the lattice circuit as compared to the ladder circuit is such that it is generally preferred for all but the narrowest bandwidth applications. The main disadvantage of the single lattice section is that it is difficult to achieve high levels of stopband attenuation because of the difficulty in maintaining the balance between the lattice arms over wide frequency ranges and also over wide operating temperature ranges. This is usually overcome by cascading a number of lattice sections.

The maximum bandwidth available from the two crystal lattice discussed above is twice the pole-zero spacing of the component crystals. For AT fundamentals it is therefore in percentage terms about 0.5%. To achieve wider bandwidths, it is common in *intermediate band* designs to introduce inductors as shown in Fig. 10.11. By the elementary network equivalences for the extraction of series and parallel elements from the arms of a lattice structure, these inductors are effectively in parallel with the crystal static capacitances. Consequently they can be used to reduce the effective values of the static capacitances and therefore increase the pole-zero spacings of the crystals and the maximum attainable bandwidth. The penalties to be paid for this increase in bandwidth are, firstly, the increased losses due to the finite

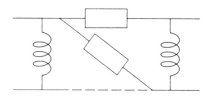

Fig. 10.11 Intermediate band lattice.

ELEMENTARY CIRCUITS: DISCRETE CRYSTAL FILTERS

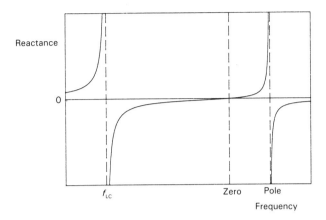

Fig. 10.12 Reactance plot for crystal with shunt inductor.

coil Q, and, secondly, the introduction of an additional pole in the frequency response of the lattice arms. This is shown in Fig. 10.12 as f_{LC} and provided the added inductance L is not too small occurs at frequencies considerably below the passband of the filter. The effect of the additional poles is to produce an unwanted passband, which can in some cases significantly affect the overall performance of the filter.

This technique by itself is only useful for achieving marginally larger bandwidths than can be achieved witout the additional inductors. It is characteristic of the two-crystal lattice as so far described that its image impedance increases from very low values at the lower edge of the passband to very high values at the upper edge. In a filter built as a cascade of crystal lattice sections with additional inductors to broaden the bandwidth, the loss resistance of the inductors appears across the signal path and has substantially more effect on the high impedance side of the passband than on the low impedance side. It is therefore typical of such filters to show a pronounced distortion of the passband, the attenuation increasing markedly from the lower to the upper passband edge.

To avoid this distortion, it is possible to choose the inductance L to resonate with the crystal static capacitances at the centre frequency of the filter. In such a case, each lattice arm impedance has a zero at the series resonance frequency of the crystal, together with two poles which are symmetrically placed either side of the zero, as shown in Fig. 10.13. The pole-zero spacing is given by

$$\Delta f/f = (C_1/C_0)^{1/2}/2 \qquad (10.2)$$

Provided the required filter bandwidth is substantially less than this spacing, the poles can, to a first approximation, be ignored, so that in the neighbourhood of the centre frequency, the lattice arm impedances can be regarded as

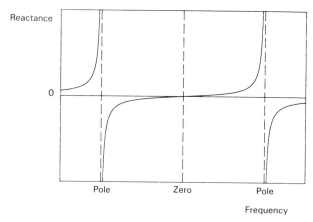

Fig. 10.13 Reactance plot for crystal with C_0 tuned out.

simple series LC circuits. Reference to Fig. 10.14 then shows that the two lattice arm reactances will have opposite signs in the frequency interval between their series resonances, so that the upper and lower edges of the filter passband will be defined by the crystal frequencies. Moreover it is clear that the image impedance $(Z_A Z_B)^{1/2}$ vanishes at both ends of the passband, giving rise to a *symmetric impedance* characteristic. Now when such lattices are cascaded, the effects of the inductor losses are spread more uniformly over the passband, leading to less relative distortion.

Although the poles of the lattice arm impedances will have little effect on the passband response if the bandwidth is sufficiently small compared to the pole-zero spacing, the poles will nevertheless affect the stopband response by

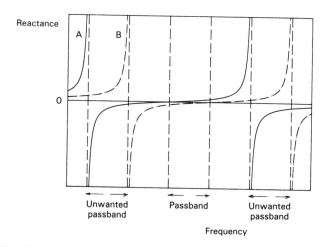

Fig. 10.14 Reactance plot for symmetric impedance lattice.

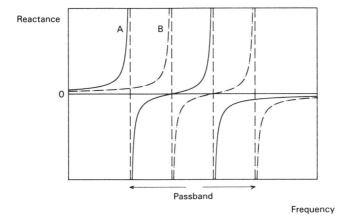

Fig. 10.15 Reactance plot for wideband lattice.

introducing unwanted passbands at the frequencies defined by Eqn (10.2). In *wideband* designs, this problem is avoided by making use of the poles to increase the bandwidth still further. This is illustrated in Fig. 10.15, from which it is clear that with the choice of pole and zero frequencies shown, the bandwidth extends from the lowest pole of one lattice impedance to the highest pole of the other lattice arm. The fractional bandwidth is then given by

$$\Delta f/f = 3(C_1/C_0)^{1/2}/2 \tag{10.3}$$

For AT fundamentals, this is approximately 10%, but it should be noted that despite the apparent attractiveness of the wideband approach, its application is restricted by the Q of available inductors to centre frequencies not much above 1 MHz.

In the lattice realizations so far discussed, the lattice arms have each contained just one crystal unit. In all cases, but particularly so in the intermediate and wideband structures, more selectivity can be obtained by utilizing the extra degrees of freedom made available by using two or more crystals in each lattice arm. This is achieved by choosing the crystal parameters in such a way that the lattice arm impedances have equal values at certain predetermined frequencies in the stopband; since the lattice is then balanced at these frequencies, attenuation peaks are thereby created.

10.3 ELEMENTARY CIRCUITS: MONOLITHIC CRYSTAL FILTERS

Since their introduction in the middle 1960s, monolithic crystal filters have become the most widely used form of crystal filter. Strictly speaking, the term

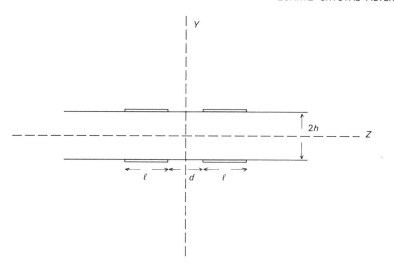

Fig. 10.16 Cross-section of monolithic dual.

monolithic refers to a filter fabricated on a single quartz plate, and incorporating two or more acoustically coupled resonators. Each resonator is formed by a pair of opposing electrodes on the major surfaces of the quartz blank, and the size, thickness, and separation of the pairs of electrodes determine the electrical response of the filter. In practice, because of manufacturing difficulties in producing sufficiently large plates, and more importantly because of the degradation of the stopband response caused by unwanted vibrational modes, it is usual to restrict the number of resonators on a plate to two only. The resulting device is termed a monolithic dual, and filters of higher degree than two are produced as a cascade of duals. The proper term for such a cascade is a 'polylithic' filter.

The operation of the monolithic dual can be explained in terms of the energy trapping theory discussed in Chapter 3. Figure 10.16 shows a cross section of a dual fabricated on a plate of thickness $2h$ in the y direction, with two pairs of parallel strip electrodes of length l along the z axis and infinite width in the x direction. The two pairs of electrodes are separated by a distance d. As in Chapter 3, thickness twist (TT) waves, propagating along the z direction, can be trapped under the electrodes as a result of the lowering of the cut-off frequency in the electroded region caused by the mass loading effect. Provided the spacing d is sufficiently large, each pair of electrodes can be regarded as constituting an independent trapped energy resonator, with there being no interaction or coupling between the two resonators. If, as shown in Fig. 10.16, the electrode lengths l are equal and if also the mass loadings are equal, then the TT mode frequencies of the two resonators will be equal. The spacing d required for there to be no interaction is determined by the rate of decay of the evanescent waves in the unelectroded regions,

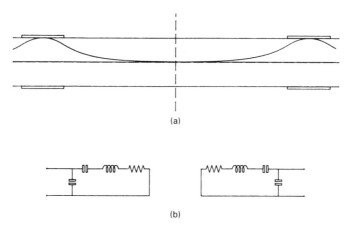

Fig. 10.17 Uncoupled resonators: (a) amplitude of vibration; (b) equivalent circuit.

specifically in the region between the electrodes. Figure 10.17(a) shows schematically the amplitude of vibration as a function of z when there is no coupling between the resonators. In this case each resonator can be represented by the conventional crystal equivalent circuit, as in Fig. 10.17(b).

As the spacing of the electrode pairs is reduced, it is clear that at some stage the evanescent wave associated with one resonator will extend into the region occupied by the other, and vice versa (Fig. 10.18a). Then by a mechanism entirely analogous to quantum mechanical 'tunnelling' through a potential barrier, energy can be transmitted from one resonator to the other, resulting in *acoustic coupling* between the resonators. In equivalent circuit terms, this coupling can be represented in various ways, one possibility being shown in Fig. 10.18(b), with a lattice equivalent in Fig. 10.18(c).

The equivalent circuit of the monolithic dual in this or alternative forms is already in one of the standard forms used in the design of narrowband coupled resonator filters. It has, in fact, precisely the form of a simple two-pole filter, with a bandwidth directly proportional to the coupling between the resonators. As already indicated, the coupling, and hence the bandwidth, is a function of the spacing d between the electrode pairs, decreasing exponentially with d. As the mass loading on the electrodes is increased, the difference in the cut-off frequencies between the electroded and the unelectroded regions increases. This results in a faster rate of decay of the evanescent waves, and consequently a decrease in the coupling and bandwidth. Thus in the manufacture of duals, both the electrode spacing and the mass loading can be adjusted to obtain the desired couplings.

A single dual has limited use as a filter, but it is straightforward to cascade duals to obtain more selective characteristics. The simplest case is that where the only additional components used are capacitors to ground at the

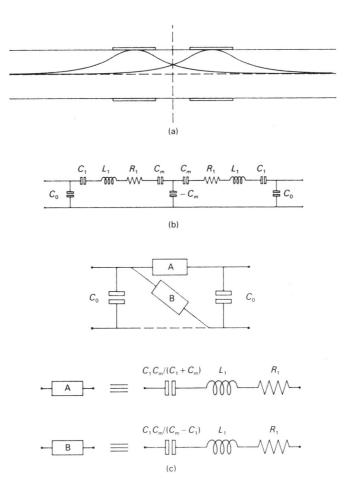

Fig. 10.18 Coupled resonators: (a) amplitude of vibration; (b) equivalent circuit; (c) lattice equivalent for monolithic dual.

junctions between adjacent duals, as in Fig. 10.19. This is the *narrowband* case, and just as for discrete crystal filters, there is an upper limit on the bandwidth achievable. For a given response type, the total capacitance required at each junction, including the static capacitances of the resonators, is inversely proportional to the bandwidth. Since the total capacitance cannot be reduced below the crystal shunt capacitances plus circuit strays without using inductors, this determines the upper limit. Just as in discrete crystal, *intermediate* band designs, the use of inductors allows wider bandwidths to be achieved, but the penalties of increased losses and some passband distortion follow. Since in practice one of the most attractive features of monolithic construction is precisely the ability to produce filters without bulky wound

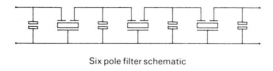

Six pole filter schematic

Circuit symbol for monolithic dual

Fig. 10.19 Six-pole narrowband polylithic filter.

components, the vast majority of monolithic or polylithic filters produced are of the narrowband type. (But note that inductors are frequently used at input and output to provide impedance matching; in some cases these are included in the filter package, but in miniature types the user has to provide the necessary inductive terminating reactance.)

The typical polylithic filter, as in Fig. 10.19, is only capable of realizing 'all-pole' filter responses in which all attenuation poles occur at infinity. Such responses include the Butterworth and Chebyshev types and various linear-phase characteristics, but not for example the 'elliptic function' type of response. It can be shown that for given (equiripple) levels of passband and stopband attenuations, and for a given number of resonators, an elliptic function filter has the sharpest cut-off of all possible filters, that is the narrowest transition region between pass and stopbands. This is achieved by the optimum placing of attenuation poles at finite frequencies in the stopband. To realize filter characteristics that have such poles using monolithic

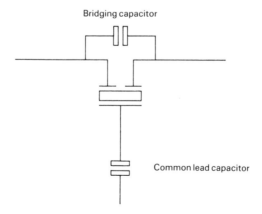

Fig. 10.20 Dual with bridging and common lead capacitors.

duals requires additional components, for example as shown in Fig. 10.20. This arrangement provides a pair of poles symmetrically disposed about the centre frequency, and can be cascaded with similar sections or standard 'all-pole' sections to give various responses. Further specialized characteristics can be achieved by adding discrete crystals to an otherwise polylithic structure, for instance to provide attenuation poles at single frequencies.

10.4 LIMITATIONS ON CRYSTAL FILTER PERFORMANCE

10.4.1 Bandwidth limitations

Crystal filters, whether discrete or monolithic, are often the only choice for narrowband applications in the range from one to several hundred megahertz. However, in specifying filters there are some fundamental limitations. It has already been pointed out that there is an upper limit on the bandwidth available, from up to 10% of the centre frequency for wideband, low frequency designs, down to fractions of a percent for designs using overtone mode crystals or duals. There are also lower limits on the bandwidth achievable.

Firstly, as the percentage bandwidth decreases, the resonator Q necessary to achieve acceptable insertion loss and passband distortion increases rapidly. For AT-cut crystals, the intrinsic material Q is approximately 10^6 at 10 MHz and is inversely proportional to frequency so that at 100 MHz it is only 10^5. Consequently the lower bandwidth limit imposed by the need for high Q is roughly proportional to frequency. Secondly, as the percentage bandwidth decreases, the problems of temperature stability and manufacturing tolerances on the resonator frequencies become more critical. If, for example, the operating temperature range of the filter is such that a frequency–temperature tolerance of ± 10 ppm on the resonator frequencies is the best that can be economically achieved, and if similarly a tolerance of ± 10 ppm is the best that can be achieved in manufacturing, then these tolerances have to be allowed for in the filter design.

For purposes of illustration, suppose the specification requirement, to be met over the operating temperature range, is for a 3 dB bandwidth of ± 2.5 kHz and a 60 dB bandwidth of 12.5 kHz at a centre frequency of 50 MHz. Since the resonator frequencies can be expected to vary a total of ± 20 ppm or ± 1 kHz due to the combined causes of temperature changes and unit to unit variations, it is necessary to allow margins of 1 kHz on either side of the passband, thus making the design bandwidth ± 3.5 kHz rather than ± 2.5 kHz, a 40% increase over the nominal. Clearly, it is not possible to decrease the *design* bandwidth indefinitely and still provide for the necessary tolerances.

The need to allow margins also has a major impact on the apparent complexity of narrowband filters. The nominal shape factor in the example above is 5:1 from 60 to 3 dB, which, as reference to any of the standard filter texts will show (for example, Zverev, 1967), is easily obtainable from a four pole Chebyshev filter with 0.5 dB ripple. However on taking into account the necessary margins on the passband, and also similar margins on the stopband, the shape factor reduces to 11.5/3.5 or approximately 3.3:1. This is *not* achievable with a four pole 0.5 dB Chebyshev filter, making either a five pole Chebyshev design or else a design with finite attenuation poles necessary. In either case, the resonator Q requirement will be more severe than in the simple four pole Chebyshev case.

10.4.2 Unwanted responses

A further major limitation on crystal filter performance is the presence of *unwanted responses* in the resonators, whether discrete or monolithic. The most troublesome responses are generally the *inharmonic* modes discussed in Chapter 3. These form a group of resonances located at frequencies just above the main or 'wanted' response. In energy-trapped designs, proper dimensioning of the electrodes and choice of mass loading can produce resonators relatively free from these unwanted or 'spurious' modes, but especially at higher frequencies and overtones the energy-trapping criteria are difficult to satisfy while still maintaining acceptable values of resonator motional inductance and resistance. Typically the mass loading cannot be reduced below a limit set by the need to maintain an electrode film thick enough to retain good electrical conductivity, and then the electrode diameters dictated by energy trapping theory turn out to be very small. As a consequence, the motional inductance, which is inversely proportional to the electrode area, is high, in conflict with the filter designer's usual need for as low an inductance as practically possible.

The presence of unwanted modes can affect the filter response in two ways. In relatively narrowband designs, the unwanted modes will typically all fall into the stopband of the filter, appearing as sharp 'spikes' that degrade the ultimate attenuation of the filter in narrow frequency intervals on the high frequency side. This is probably the least troublesome manifestation of 'spurious' modes. As the bandwidth increases, the filter impedance will generally increase, and the unwanted modes will both increase in level, and appear relatively closer to the centre frequency, until they actually occur in the transition region. Further increase in bandwidth finally leads to a situation where the spurious responses begin to appear in the passband of the filter, showing up as sharp notches superimposed on an otherwise smooth response. The occurrence of spurious modes in the passband can very often limit the

10.4.3 Group delay characteristics

With the rapid increase in data transmission over radio communications networks, the group delay performance of crystal filters is becoming increasingly important. As mentioned previously in Section 10.1, for the class of filters known as *minimum-phase* filters, the delay and amplitude characteristics are not independent. For such filters, high selectivity implies rapidly changing group delay at the passband edges, whereas, flat delay characteristics imply poor selectivity. Nearly all the crystal filter realizations currently used fall into this class, and so cannot satisfy independent specifications on delay and amplitude. This situation is in contrast to that existing for SAW filters, where generally the phase and amplitude characteristics can be independently controlled. However, crystal filters have been developed using unconventional realization techniques which can satisfy simultaneous phase and amplitude specifications and these show promise for the future.

10.4.4 Non-linear effects

Crystal filters containing such known nonlinear components as ferrite cores can be expected to show the same effects as other devices using the same components. Hence ferrites and other nonlinear materials should be avoided in the manufacture of filters where intermodulation or other amplitude dependent phenomena are to be avoided. Assuming this to be the case, the residual non-linearities in crystal filters can generally be assigned to the individual resonators. As such, a fuller discussion is given in Chapter 7, but it can be pointed out here that in the case of intermodulation, there appear to be two separate mechanisms acting according to whether the test tones applied are in the passband or in the stopband.

To be specific, consider two test tones at frequencies $f_0 + df$ and $f_0 + 2df$, with f_0 being the centre frequency of the filter. Third-order intermodulation then gives rise to an intermodulation product (IMP) at the centre frequency f_0. If the frequency offset df is such that both test signals are in the filter stopband, then the resonator currents can be expected to be small, since the crystals are being driven at frequencies far from resonance. On the other hand, if df is such that both signals fall within the filter passband, then the resonator currents can be expected to be much larger as the test frequencies will then be closer to the crystal resonance frequencies.

In the latter case, it appears that the intermodulation products are the result of intrinsic nonlinearities in the resonator material, specifically in the elastic stress-strain relationship. This case has been treated theoretically by Tiersten (1974, 1975), with the result that the intermodulation ratio, that is the ratio of the input power in the test tones to the power delivered to the load at the intermodulation frequency, turns out to be proportional to the square of the overtone order and to the square of the electrode area (Smythe, 1974). That is for the in-band case, filters using overtone resonators are generally to be preferred to those using fundamental mode resonators.

In the case where the test tones are in the stopband, the intermodulation mechanism has been found to be largely process dependent. The key factors appear to be the same as those discussed in Chapter 7 in relation to the increase in crystal resistance and the change in resonator frequency sometimes observed at low drive levels, that is surface finish and cleanliness and the quality of the electrode films. Consequently, this type of intermodulation is difficult to quantify, and may well not depend on the test tone power levels in the expected way, often being relatively worse than expected at low power levels and better than expected at higher power levels.

Other non-linear effects that may be observed in crystal filters at low input power levels can also be associated with the process factors mentioned in the previous paragraph. At very high power levels other effects can be observed that result from shifts in resonator frequency and resistance due to the generation of thermal gradients and stresses in the crystal blanks, but such power levels should be avoided.

Appendix 1
Explanation of piezoelectric effect

The first explanation of the origin of the piezoelectric effect in terms of molecular structure was given by Lord Kelvin soon after the Curies' original discovery (Heising, 1946). Although the structure assumed by Kelvin has since been proved incorrect by X-ray crystallography, his explanation remains useful in a qualitative sense.

Figure A1.1 shows an arrangement of positive and negative ions which can serve as a crude model of the unit cell of a quartz crystal in the plane normal to the optic axis. The six ions are located at the corners of a regular hexagon. Assuming that the ions have charges $+q$ and $-q$, the net charge in the unit cell is zero. Moreover, because of the alternate arrangement of the positive and negative ions, the net dipole moment is also zero.

If now it is supposed that during a deformation of the crystal caused by the application of external stresses, the mutual forces acting between the ions in

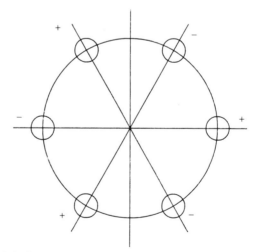

Fig. A1.1 Model of quartz 'molecule'.

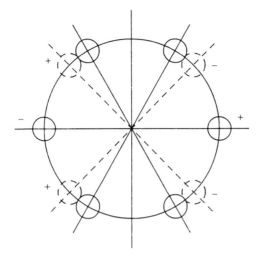

Fig. A1.2 Vertical compressional strain.

the cell are such as to maintain both the distance apart of opposite pairs of ions, and also their collinearity with the cell centre, then the deformation of the unit cell can be described in terms of the rotations of the lines joining the opposite pairs of ions. Figure A1.2 illustrates the case of a compression in the vertical direction. The ion pair in the horizontal direction is unaltered, but the remaining two ion pairs are rotated towards the horizontal. Clearly the centre of gravity of the positive ions is shifted towards the left, whereas that of the negative ions is shifted towards the right, the combined effect being to produce a non-zero dipole moment in the horizontal direction.

The reverse situation of a tensile stress in the vertical direction again leaves the horizontal ion pair unchanged but rotates the other two pairs away from the horizontal, producing a dipole moment oppositely directed to that produced by compression. Assuming a homogeneous strain, summing the dipole moments of all the unit cells throughout the material therefore results in an electrical polarization that reverses with the strain, that is, *piezoelectricity*. In this simple model, the horizontal direction lies along an *electric axis*. By inspection, there are three such axes, so that to this extent the model reflects the trigonal symmetry of quartz.

The same model can be used to demonstrate the production of electrical polarization by longitudinal strains in the horizontal direction, and also by shear strains in the plane of the paper. In the former case, the polarization is again along the horizontal axis, but a shear strain produces a polarization along the vertical axis. The details are however hardly relevant in practice since the actual structure of quartz is considerably more complicated than suggested here (cf. Vigoreux and Booth, 1950; Brice, 1985).

Appendix 2
Vectors and tensors

A2.1 COORDINATE SYSTEMS AND TRANSFORMATIONS

The position of a point P in a three-dimensional Euclidean space can be specified by its coordinates (x, y, z) in a rectangular cartesian coordinate system $Oxyz$ with origin O and three mutually perpendicular axes Ox, Oy and Oz. The distance OP from the origin to the point P is given by the three-dimensional generalization of Pythagoras' theorem in plane Euclidean geometry

$$OP = (x^2 + y^2 + z^2)^{1/2}$$

If n_x, n_y and n_z are the cosines of the angles made by OP with the coordinate axes Ox, Oy and Oz respectively, then

$$x = n_x OP \qquad y = n_y OP \qquad z = n_z OP$$

Thus the rectangular coordinates of any point Q on a line drawn through O and P are proportional to the length OQ.

The choice of coordinate system is initially arbitrary, that is any other rectangular cartesian system would serve as well as the system $Oxyz$. Suppose $Ox'y'z'$ is such another system with the same origin O but mutually perpendicular axes Ox', Oy' and Oz', and suppose the point P has coordinates (x', y', z') in this system. Then by the argument already used, the primed coordinates of any point Q on a line drawn through O and P are also proportional to the length OQ. If Q is such that $OQ = kOP$ then the unprimed and primed coordinates of Q will be (kx, ky, kz) and (kx', ky', kz'), respectively. This implies a linear relationship between the primed and unprimed coordinates that can be written

$$x = a_{xx} \cdot x' + a_{xy} \cdot y' + a_{xz} \cdot z'$$
$$y = a_{yx} \cdot x' + a_{yy} \cdot y' + a_{yz} \cdot z'$$
$$z = a_{zx} \cdot x' + a_{zy} \cdot y' + a_{zz} \cdot z'$$

This and similar relations can be written in much more compact form by relabelling the coordinate axes Ox_1, Ox_2 and Ox_3 instead of Ox, Oy and Oz, and writing the coordinates of P as (x_1, x_2, x_3) instead of (x, y, z). Then the relation between primed and unprimed coordinates can simply be written

$$x_i = a_{ik}x_k' \tag{A2.1}$$

where it is understood that the suffices i, k have the range 1, 2, 3 and that a repeated suffix implies summation over the range of the suffix (Einstein summation convention). Thus, when written out in full,

$$x_i = a_{i1}x_1' + a_{i2}x_2' + a_{i3}x_3' \tag{A2.2}$$

In this notation the square of the length OP is simply written

$$OP^2 = x_i x_i = x_i' x_i' \tag{A2.3}$$

since the length is independent of the coordinate system. This implies a set of conditions on the coefficients a_{ik}, namely

$$a_{ik}a_{il} = \delta_{kl} (= 1 \text{ if } k = l, = 0 \text{ if } k <> l) \tag{A2.4}$$

where δ_{kl} is the Kronecker delta. Using these conditions the equations (A2.1) can be inverted to give the primed coordinates in terms of the unprimed

$$x_i' = a_{ki}x_k \tag{A2.5}$$

In general a coordinate transformation described by Eqns (A2.1) and (A2.4) is termed an orthogonal transformation and is physically represented by either a rigid rotation of the coordinate system, or a rotation together with an inversion, that is a reversal in sense of one of the axes leading to a change from a right-handed to a left-handed system or vice-versa.

A2.2 SCALARS, VECTORS AND TENSORS

A *scalar* quantity is one completely specified by its measure in some previously defined unit. A scalar quantity such as mass, charge or temperature is *invariant* under coordinate transformations.

The physical displacement OP is the prototype *vector* quantity, having both magnitude and direction and being completely specified by the three coordinates x_i of P relative to O in a rectangular coordinate system Ox_i. The x_i are termed the *components* of the displacement vector OP and under an orthogonal transformation of coordinates the components transform according to Eqn (A2.1). In general a vector quantity is defined as any set of three numbers v_i associated with a rectangular cartesian coordinate system Ox_i which transform under an orthogonal transformation in the same way as the components of a physical displacement. The v_i are the *components* of the vector in the coordinate system Ox_i. If v_i' are the components in another coordinate system Ox_i', then *by definition*

$$v_i' = a_{ki}v_k \tag{A2.6}$$

It follows from the kinematic definitions of velocity and acceleration as time derivatives of displacement that they are vector quantities. The product of a vector and a scalar is also a vector quantity, hence momentum and force are also vectors. Since electric charge is a scalar, then it also follows that the electric field is a vector.

If v_i and w_i are two vectors, their *scalar product* or *dot product* is defined as the sum $v_i w_i$. From Eqns (A2.4) and (A2.6) the scalar nature of the dot product follows immediately

$$v_i w_i = a_{ik} a_{il} v_k' w_l' = \delta_{kl} v_k' w_l' = v_k' w_k' \tag{A2.7}$$

The *tensor product* of the vectors v_i and w_i is defined as the set of nine quantities $v_i w_k$. The transformation law of the tensor product follows directly from Eqn (A2.6) and is

$$v_i' w_k' = a_{ji} a_{lk} v_j w_l \tag{A2.8}$$

This rule is taken to be the defining property of a *second rank tensor* quantity. Thus a second rank tensor is any collection of nine quantities t_{ik} associated with a rectangular cartesian coordinate system which transform under an orthogonal coordinate transformation according to

$$t_{ik}' = a_{ji} a_{lk} t_{jl} \tag{A2.9}$$

The t_{ik} are the components of the tensor in the coordinate system Ox_i.

Higher rank tensors are defined by straightforward generalization of Eqn (A2.9). Thus a third rank tensor has $3^3 = 27$ components t_{ikm} and a fourth rank tensor $3^4 = 81$ components t_{ikmp}, with transformation laws given by Eqns (A2.10) and (A2.11)

$$t_{ikm}' = a_{ji} a_{lk} a_{nm} t_{jln} \tag{A2.10}$$

$$t_{ikmp}' = a_{ji} a_{lk} a_{nm} a_{qp} t_{jlnq} \tag{A2.11}$$

Any second rank tensor t_{ik} can be written as the sum of a symmetric and an antisymmetric part

$$t_{ik} = t_{(ik)} + t_{[ik]}$$

where

$$t_{(ik)} = (t_{ik} + t_{ki})/2$$
$$t_{[ik]} = (t_{ik} - t_{ki})/2$$

For the symmetric part, $t_{(ik)} = t_{(ki)}$, whereas for the antisymmetric part $t_{[ik]} = -t_{[ki]}$. The antisymmetric part thus has just three independent components $t_{[23]}$, $t_{[31]}$, and $t_{[12]}$. These can be written

$$v_i = \tfrac{1}{2}\epsilon_{ikm} t_{[km]} = \tfrac{1}{2}\epsilon_{ikm} t_{km} \tag{A2.12}$$

where ϵ_{ikm} is the permutation symbol, equal to $+1$ if ikm is an even permutation of 123, -1 if ikm is an odd permutation of 123, and 0 if any two of i, k, m are equal. The transformation properties of the v_i can be determined by writing the corresponding expression in the primed coordinate system

$$v_j' = \tfrac{1}{2}\epsilon_{jln} t_{[ln]}' = \tfrac{1}{2}\epsilon_{jln} a_{kl} a_{mn} t_{[km]}$$

Therefore

$$a_{ij} v_j' = \tfrac{1}{2}\epsilon_{jln} a_{ij} a_{kl} a_{mn} t_{[km]}$$

But from the definition of a determinant

$$\epsilon_{ikm} \det\{a_{pq}\} = \epsilon_{jln} a_{ij} a_{kl} a_{mn}$$

so that

$$\begin{aligned} v_i &= \det^{-1}\{a_{pq}\}\, a_{ij} v_j' \\ v_j' &= \det\{a_{pq}\}\, a_{ij} v_i \end{aligned} \qquad (A2.13)$$

Comparing Eqns (A2.13) with (A2.6) shows that the v_i transform as a vector except for the factor $\det\{a_{pq}\}$. From Eqn (A2.4) it follows that the determinant can only have the values $+1$ or -1, since its square must be $+1$. It is also clear that since the determinant is a continuous function of the a_{pq} and must have value $+1$ for the identity transformation $a_{pq} = \delta_{pq}$, it must have the value $+1$ for all transformations obtained by a continuous variation of the a_{pq} from δ_{pq}. Physically, this means that all orthogonal transformations corresponding to a rotation of the coordinate system have $\det\{a_{pq}\} = +1$, whereas transformations that involve a change in sense of the coordinate system have $\det\{a_{pq}\} = -1$. The former transformations are termed *proper*, the latter *improper*.

Equations (A2.13) then show that the v_i transform as a vector under proper orthogonal transformations, but change sign relative to a true vector under improper transformations. Generally, Eqns (A2.13) are used as the defining property for a class of physical quantities known as *axial* or *pseudo-* vectors. The mathematical term for quantities such as v_i is a *relative vector of weight +1*. By extension, a relative tensor of weight W obeys by definition a transformation law

$$t_{jln..}' = \det\{a_{pq}\}^W a_{ij} a_{kl} a_{mn} \cdots t_{ikm..} \qquad (A2.14)$$

Of particular interest is the relative vector u_i associated with the antisymmetric part of the tensor product of two vectors v_i and w_i

$$u_i = \epsilon_{ikm} v_{[k} w_{m]} = \epsilon_{ikm} v_k w_m \qquad (A2.15)$$

This is the *vector product* or *cross product* of v_i and w_i. Equation (A2.15) shows clearly that under an inversion of the coordinate axes in which the components of a true or absolute vector change sign, the components of the axial vector u_i are unaltered.

A2.3 LENGTHS, ANGLES, AREAS AND VOLUMES

Suppose that $x_i^{(1)}$ are the components of the displacement vector from the origin O to a point $P^{(1)}$. Similarly, let $x_i^{(2)}$, $x_i^{(3)}$ be the components of the displacements from O to points $P^{(2)}$ and $P^{(3)}$, respectively. Then the length or magnitude L of the displacement $OP^{(1)}$ is given by

$$L^2 = \delta_{kl}x_k^{(1)}x_l^{(1)} = x_k^{(1)}x_k^{(1)} \tag{A2.16}$$

If the angle between $OP^{(1)}$ and $OP^{(2)}$ is β and the lengths of $OP^{(1)}$ and $OP^{(2)}$ are $L_{(1)}$ and $L_{(2)}$, respectively, then

$$x_i^{(1)}x_i^{(2)} = L_{(1)}L_{(2)}\cos(\beta)$$

In particular, if $OP^{(1)}$ and $OP^{(2)}$ are perpendicular, then $x_i^{(1)}x_i^{(2)} = 0$.

Now consider the parallelogram with adjacent edges $OP^{(1)}$ and $OP^{(2)}$. Let A_{23} be the area of the projection of the parallelogram on the Ox_2x_3 coordinate plane. Then A_{23} is given by

$$A_{23} = x_2^{(1)}x_3^{(2)} - x_3^{(1)}x_2^{(2)} = 2x_{[2}^{(1)}x_{3]}^{(2)}$$

Cyclic permutation of the indices gives the projections A_{31} and A_{12} on the remaining coordinate planes. The quantities A_{ik} form the components of an antisymmetric tensor

$$A_{ik} = 2x_{[i}^{(1)}x_{k]}^{(2)}$$

In analogy with Eqn (A2.12), a relative vector of weight $+1$ can be associated with A_{ik} by

$$A_i = \tfrac{1}{2}\epsilon_{ikm}A_{km} = \epsilon_{ikm}x_k^{(1)}x_m^{(2)} \tag{A2.17}$$

A_i is the vector area of the parallelogram with adjacent edges $OP^{(1)}$ and $OP^{(2)}$. From Eqn (A2.17) it follows that $A_i x_i^{(1)} = A_i x_i^{(2)} = 0$, so that A_i is perpendicular to both $x_i^{(1)}$ and $x_i^{(2)}$. It also follows that the sign of A_i depends on the order in which the edge vectors are taken, that is, the area has two definite orientations.

Finally, consider the parallelepiped with adjacent edges $OP^{(1)}$, $OP^{(2)}$ and $OP^{(3)}$. The volume V of the parallelepiped is given by the scalar product of one of the edge vectors with the vector area of the parallelogram defined by the other two. Therefore

$$V = x_i^{(1)}\epsilon_{ikm}x_k^{(2)}x_m^{(3)} = \epsilon_{ikm}x_i^{(1)}x_k^{(2)}x_m^{(3)} \tag{A2.18}$$

Again the sign of V depends on the order in which the edge vectors are taken, so that V has two possible orientations. Under an orthogonal transformation, V transforms according to

$$V' = \epsilon_{jln}x_j^{(1)'}x_l^{(2)'}x_n^{(3)'} = \epsilon_{jln}a_{ij}a_{kl}a_{mn}x_i^{(1)}x_k^{(2)}x_m^{(3)}$$

Therefore

$$V' = \det\{a_{pq}\}\, \epsilon_{ikm} x_i^{(1)} x_k^{(2)} x_m^{(3)} = \det\{a_{pq}\}\, V \qquad (A2.19)$$

V is thus a relative scalar of weight $+1$.

A2.4 VECTOR AND TENSOR FIELDS

The mass or charge associated with a particle depends only on the particle and, at least in classical mechanics, is independent of the position of the particle. Other physical quantities are however defined at each point in a region of space and over an interval of time and have to be expressed mathematically as functions of position and time. Such quantities are known as *fields*. A *scalar field* F is represented by a single function of position x_i and time t, so that $F = F(x_i, t)$. By definition the value of the field F at a given point is invariant under coordinate transformations, although the form of its functional dependence on the coordinates will be course depend on the coordinate system chosen. Thus if $x_i = a_{ik} x_k'$ is an orthogonal transformation

$$F = F(x_i, t) = F(a_{ik} x_k', t) = F'(x_i', t)$$

An example of a scalar field is the temperature at each point of a continuum, $T = T(x_i, t)$.

Vector and tensor fields are by extension sets of functions $t_{ik\ldots}(x_r, t)$ defined over a given space-time region in such a way that the values of the functions at a given space-time point transform according to the appropriate law under a coordinate transformation

$$t_{ik\ldots}(x_r, t) = a_{ij} a_{kl} \cdots t_{jl\ldots}'(x_r, t)$$

Clearly both the value and the functional form of a component of a vector or tensor field depends on the coordinate system chosen.

Let F be a scalar field $F = F(x_i, t)$. Define $F_{,i}$ to be the partial derivative of F with respect to x_i. Then the $F_{,i}$ form the components of a vector field known as the *gradient* of the scalar field F. This follows from the elementary rules of calculus

$$F_{,i} = F_{,k}' x_{k,i}' = a_{ik} F_{,k}' \qquad (A2.20)$$

where $F_{,k}'$ is the partial derivative of F with respect to x_k'. Equation (A2.20) shows that the gradient $F_{,i}$ satisfies the defining rule (A2.6) for a vector quantity.

Now let F_i be a vector field. The sum $F_{i,i} = F_{1,1} + F_{2,2} + F_{3,3}$ is a scalar field known as the *divergence* of F_i. Its scalar nature follows directly from

$$F_{i,i} = (a_{ik} F_k')_{,j} x_{j,i}' = a_{ik} a_{ij} F_{k,j}' = F_{k,k}'$$

where the orthogonality condition Eqn (A2.4) has been used.

In general the partial derivatives $F_{i,k}$ of a vector field form the components of a second rank tensor. The antisymmetric part $F_{[i,k]}$ has just three independent components $F_{[2,3]}$, $F_{[3,1]}$ and $F_{[1,2]}$, which can be written

$$v_i = \epsilon_{ikm} F_{[k,m]} = \epsilon_{ikm} F_{k,m} \tag{A2.21}$$

The v_i transform as a relative vector of weight 1 known as the *curl* of the vector field F_i.

The importance of the divergence and curl of a vector field F_i lies in their appearance in the integral theorems of Gauss and Stokes. First let V be a region in space bounded by the closed surface S. At each point of S let n_i be the outward-pointing unit normal (ie, a vector of unit length drawn perpendicular to S and pointing out of V). Then if F_i is a vector field defined over V, Gauss's theorem states that

$$\int_V F_{i,i} \, dV = \oint_S F_i n_i \, dS \tag{A2.22}$$

The restriction of F_i to a vector field is not necessary, and Gauss's theorem may be stated for any tensor field $t_{ik..p}$

$$\int_V t_{ik..p,p} \, dV = \oint_S t_{ik..p} n_p \, dS \tag{A2.23}$$

Gauss's theorem is frequently referred to as the divergence theorem.

If now S is a surface bounded by a closed loop C, and if the unit normal n_i on S and the positive sense of C form a right-handed screw, Stokes's theorem states that

$$\int_S \epsilon_{ikm} n_i F_{k,m} \, dS = \oint_C F_i \, dl_i \tag{A2.24}$$

where the dl_i are the components of an element of length along the boundary curve C.

A2.5 CHANGE OF VARIABLES IN A MULTIPLE INTEGRAL

In continuum mechanics it is frequently convenient in applications of the divergence theorem to make use of the rule for change of variables in a multiple integral. For completeness, this rule is stated here.

Let f be a function of the n real variables x_i, where i now runs from 1 to n. The x_i can be regarded as the coordinates of a point in an n-dimensional space, x-space. Let X_i be a second set of n variables, which can similarly be regarded as the coordinates of a point in an n-dimensional X-space. Suppose now that the x_i and the X_i are connected by a set of functions $x_i = x_i(X_k)$ and their inverses $X_i = X_i(x_k)$. The necessary and sufficient condition that the inverses should exist is that the *Jacobian determinant* J should not vanish in the region of interest

$$J = \det\{x_{i,k}\} = 1/\det\{X_{i,k}\} \neq 0$$

If f is defined in a region V of x-space and V' is the region in X-space into which V is mapped by the functions $X_i(x_k)$, then the integral of f over V can be transformed into an integral over V' by

$$\int_V f(x_i)\,dx_1\,dx_2\ldots dx_n = \int_{V'} f(x_i(X_k))\,J\,dX_1\,dX_2\ldots dX_n$$

or in alternative notation

$$\int_V f(x_i)\,dV = \int_{V'} f(x_i(X_k))\,J\,dV' \tag{A2.25}$$

A2.6 EIGENVALUES OF A REAL SYMMETRIC TENSOR

For any second rank tensor t with components t_{km} and any vector v with components v_m, the products $w_k = t_{km}v_m$ form the components of a second vector w, which can be described as the result of the tensor t *operating* on the vector v. The *eigenvalue problem* for the tensor t is the problem of determining all vectors e and scalars E such that

$$t_{km}e_m = Ee_k \tag{A2.26}$$

that is, such that the effect of t operating on e is to leave the direction of e unchanged. Vectors e satisfying Eqn (A2.26) are *eigenvectors* of t, and the scalars E are the corresponding *eigenvalues*.

Equation (A2.26) can be written as

$$(t_{km} - E\delta_{km})e_m = 0 \tag{A2.27}$$

which is a system of homogeneous linear equations for the components e_m. The condition for non-trivial solutions to exist is that the determinant of coefficients should vanish

$$\det\{t_{km} - E\delta_{km}\} = 0$$

Expansion of the determinant leads to a cubic equation in E, which in general will have three (complex) roots $E^{(1)}$, $E^{(2)}$ and $E^{(3)}$. For each eigenvalue $E^{(k)}$, Eqns (A2.27) can be solved for the (complex) components $e_m^{(k)}$ of the associated eigenvector. (Note that since the eigenvectors are only determined up to a scalar multiple, they can be assumed to be normalized to unit length through $e_m e_m^* = 1$, where the * indicates the complex conjugate.)

Provided the components t_{km} of t are real, the coefficients of the cubic in E will all be real, and it then follows that there will be at least one real eigenvalue, and an associated real eigenvector. If in addition to being real, the t_{km} are also symmetric, so that $t_{km} = t_{mk}$, then it can be proved that all the eigenvalues and eigenvectors are real. For if E and e_m satisfy

$$t_{km}e_m = Ee_k \tag{A2.28}$$

then by taking complex conjugates it follows that

$$t_{km}e_m^* = E^*e_k^* \tag{A2.29}$$

Multiplying Eqns (A2.28) and (A2.29) by e_k^* and e_k, respectively, and subtracting leads to

$$e_k^*t_{km}e_m - e_k t_{km}e_m^* = (E - E^*)e_k e_k^* \tag{A2.30}$$

But by the symmetry of t_{km} the left side of Eqn (A2.30) vanishes and hence $E = E^*$, that is, E is real. It then follows from Eqn (A2.27) that the associated eigenvector e_k is also real.

If now E and e_k, F and f_k are two eigenvalues and associated eigenvectors, it follows that

$$f_k t_{km} e_m = E f_k e_k$$
$$e_k t_{km} f_m = F f_k e_k$$

Again by the symmetry of t_{km} the left sides are equal so that subtracting leads to

$$(E - F) f_k e_k = 0 \tag{A2.31}$$

Hence if E and F are not equal it follows that the associated eigenvectors are orthogonal. If $E = F$ this does not follow, but in such a case it is easily shown that any linear combination of e_k and f_k is also an eigenvector of t_{km} corresponding to the same eigenvalue $E = F$. Thus it is always possible to choose eigenvectors such that $e_k f_k = 0$.

In summary a real symmetric tensor t has three real eigenvalues. In the case where the eigenvalues are distinct, the associated eigenvectors are necessarily mutually orthogonal, whereas in the degenerate case where two or all three eigenvalues are equal, the eigenvectors may be chosen to be orthogonal.

Supposing that three orthonormal eigenvectors $e_m^{(k)}$ have been selected, it is then possible to use these as the basis for a new coordinate system, that is a coordinate system Ox_k' in which the eigenvectors have components $(1, 0, 0)$, $(0, 1, 0)$ and $(0, 0, 1)$. If the coordinate transformation is described by $x_k = a_{km}x_m'$ as in Eqn (A2.1), then clearly $e_m^{(k)} = a_{mk}$. The components of t in the new coordinate system are given via Eqns (A2.9) as

$$t_{ik}' = a_{ji}a_{lk}t_{jl} = e_j^{(i)} e_l^{(k)} t_{jl}$$

But since the $e_m^{(k)}$ are eigenvectors, this reduces to

$$t_{ik}' = E^{(k)} \delta_{ik} \quad \text{(no sum on } k\text{)} \tag{A2.32}$$

That is, in the coordinate system defined by the eigenvectors, the components t_{km}' vanish if $k <> m$, and the diagonal entries t_{11}', t_{22}' and t_{33}' are just the eigenvalues $E^{(1)}$, $E^{(2)}$ and $E^{(3)}$.

Finally, consider the scalar quantity $Q = t_{km}x_k x_m$. In the eigenvector coordinate system, Q takes the simple form

$$Q = E^{(1)}x_1^2 + E^{(2)}x_2^2 + E^{(3)}x_3^2$$

If all the eigenvalues are positive, it then follows that $Q > 0$ for all values of x_k except $x_k = 0$, when $Q = 0$. When this is the case, the real symmetric tensor t is described as *positive-definite*. Reversing the argument, if it can be shown that $Q > 0$ for all x_k not equal to zero, then it follows that all the eigenvalues of the associated real, symmetric tensor are real and *positive*. (This result is of fundamental importance in the theory of wave propagation in anisotropic materials, discussed in Chapter 2.)

Appendix 3
Continuum mechanics

A3.1 DEFORMATION AND STRAIN

Let B be a body that occupies a volume V_0 bounded by a surface S_0 at a time t_0. Let X be a point of B which at time t_0 has coordinates X_K with respect to a rectangular cartesian coordinate system Ox_K. Then the X_K can be regarded as labelling the point X. A motion of the body B can be described by giving the coordinates x_k of X at time $t > t_0$ as functions f_k of X_K and t

$$x_k = f_k(X_K, t) \tag{A3.1}$$

It is assumed that the functions f_k can be inverted to give an alternative description of the motion in the form

$$X_K = F_K(x_k, t) \tag{A3.2}$$

The necessary and sufficient condition for the inverse functions F_K to exist is that the Jacobian determinant $J = \det\{\partial f_k/\partial X_K\}$ be non-zero in V_0.

From Eqn (A3.1) follows

$$dx_k = (\partial f_k/\partial X_K) dX_K = x_{k,K} dX_K \tag{A3.3}$$

where the notation $x_{k,K} = \partial f_k/\partial X_K$ is used. Similarly $X_{K,k}$ may be written for $\partial F_K/\partial x_k$, and it then follows by the chain rule for differentiation that

$$x_{k,K} X_{K,l} = \delta_{kl} \quad \text{and} \quad X_{K,k} x_{k,L} = \delta_{KL} \tag{A3.4}$$

Taking determinants in Eqn (A3.4) leads to

$$\det\{x_{k,K}\} \det\{X_{K,l}\} = 1$$

so that the non-vanishing of the Jacobian $J = \det\{x_{k,K}\}$ implies the non-vanishing of the Jacobian $J^{-1} = \det\{X_{K,k}\}$ of the inverse functions F_K in Eqn (A3.2).

Equation (A3.3) shows how an elementary vector dX_K joining neighbouring points of B is transformed into a vector dx_k by the motion. If dL and dl are the magnitudes of dX_K and dx_k, respectively, then $dL^2 = dX_K dX_K$ and $dl^2 = dx_k dx_k$ and by Eqn (A3.3)

$$dl^2 = x_{k,K} x_{k,L} dX_K dX_L = C_{KL} dX_K dX_L \tag{A3.5}$$

DEFORMATION AND STRAIN 173

where

$$C_{KL} = x_{k,K} x_{k,L} \tag{A3.6}$$

The C_{KL} form the components of a symmetric second rank tensor known as *Green's deformation tensor*. In a motion such that $C_{KL} = \delta_{KL}$ the distance between pairs of neighbouring points does not change, since

$$dl^2 = C_{KL} dX_K dX_L = \delta_{KL} dX_K dX_L = dL^2$$

Hence B moves as a rigid body when $C_{KL} = \delta_{KL}$.

Generally, the difference in the squared lengths dl^2 and dL^2 is

$$dl^2 - dL^2 = (C_{KL} - \delta_{KL}) dX_K dX_L = 2 S_{KL} dX_K dX_L \tag{A3.7}$$

where

$$S_{KL} = (C_{KL} - \delta_{KL})/2 \tag{A3.8}$$

The S_{KL} are the components of the symmetric *Lagrangian finite strain tensor*, which vanishes in a rigid body motion.

Now let $dX_K^{(1)}$ and $dX_K^{(2)}$ be two elementary vectors defining an element of area dA_{KL} in B (cf. Section A2.3). The dA_{KL} are given by

$$dA_{KL} = 2 \, dX_{[K}^{(1)} dX_{L]}^{(2)}$$

The element of area is transformed by the motion into the element da_{kl}, where

$$da_{kl} = 2 \, dx_{[k}^{(1)} dx_{l]}^{(2)}$$

and $dx_k^{(1)}$, $dx_k^{(2)}$ are related to $dX_K^{(1)}$, $dX_K^{(2)}$ by Eqn (A3.3). Therefore

$$da_{kl} = x_{k,K} x_{l,L} dA_{KL} \tag{A3.8}$$

In terms of the associated vector elements of area da_k and dA_K defined by $da_k = \tfrac{1}{2}\epsilon_{klm} da_{lm}$ and $dA_K = \tfrac{1}{2}\epsilon_{KLM} dA_{LM}$ (Eqn (A2.17))

$$da_k = J X_{K,k} dA_K \tag{A3.9}$$

If $dX_K^{(3)}$ is a third elementary vector, then $dX_K^{(1)}$, $dX_K^{(2)}$ and $dX_K^{(3)}$ together define an element of volume dV in B given by (Eqn (A2.18))

$$dV = \epsilon_{KMP} dX_K^{(1)} dX_M^{(2)} dX_P^{(3)}$$

which is transformed in the motion to a volume element dv, where

$$dv = J \, dV \tag{A3.10}$$

In summary, Eqns (A3.1), (A3.3), (A3.9) and (A3.10) show how points, line elements, area elements and volume elements are transformed by the motion of B.

A3.2 KINEMATICS AND CONSERVATION OF MASS

From Eqn (A3.1) and velocity v_k of the particle X is just the partial derivative of $f_k(X_K, t)$ with respect to t, keeping X_K constant. As such v_k is to be regarded as a function of X_K and t. However, v_k can equally well be regarded as a function of x_k and t by substituting $X_K(x_k, t)$ for X_K. Thus

$$v_k = v_k(X_K, t) = v_k(X_K(x_k, t), t) = v_k(x_k, t)$$

Here v_k is being used to denote two distinct functions along with their values. To avoid potential confusion the following conventions are adopted:

(a) the partial time derivative of v_k when regarded as a function of X_K and t is indicated by a superimposed dot and termed the *material time derivative*;
(b) the partial time derivative of v_k when regarded as a function of x_k and t is written $\partial v_k/\partial t$ in the conventional manner;
(c) the notation $v_{k,l}$ is used to indicate the partial derivative of v_k with respect to x_l with v_k being regarded as a function of x_k and t;
(d) the notation $v_{k,K}$ is used to indicate the partial derivative of v_k with respect to X_K with v_k being regarded as a function of X_K and t.

It follows that the material time derivative of v_k is

$$\dot{v}_k = v_l v_{k,l} + \partial v_k/\partial t \tag{A3.11}$$

If the same conventions are applied to any other field quantity $f = f(x_k, t) = f(x_k(X_K, t), t) = f(X_K, t)$, then it also follows that

$$\dot{f} = v_l f_{,l} + \partial f/\partial t \tag{A3.12}$$

Taking the material time derivative of Eqn (A3.3) gives

$$\overline{dx_k} = \overline{x_{k,K}} dX_K \tag{A3.13}$$

By changing the order of differentiation, $\overline{x_{k,K}} = v_{k,K}$, and by the chain rule, $v_{k,K} = v_{k,l} x_{l,K}$. Therefore $\overline{dx_k} = v_{k,l} x_{l,K} dX_K$ and by Eqn (A3.3)

$$\overline{dx_k} = v_{k,l} dx_l \tag{A3.14}$$

Differentiating Eqn (A3.6), the material time derivative of C_{KL} is

$$\dot{C}_{KL} = 2x_{k,K} x_{l,L} v_{(k,l)} \quad (= 2\dot{S}_{KL}) \tag{A3.15}$$

Therefore from Eqn (A3.5)

$$\overline{dl^2} = 2v_{(k,l)} dx_k dx_l \tag{A3.16}$$

The Jacobian $J = \det\{x_{k,K}\}$ can be written

$$J = (1/3!)\epsilon_{kmp}\epsilon_{KMP} x_{k,K} x_{m,M} x_{p,P}$$

Differentiating and using the result $v_{k,K} = v_{k,l}x_{l,K}$ leads directly to

$$\dot{J} = J v_{k,k} \tag{A3.17}$$

Differentiation of the identity $X_{K,k}x_{k,L} = \delta_{KL}$ leads to

$$\overline{X_{K,k}} = -X_{K,l}v_{l,k} \tag{A3.18}$$

Equations (A3.17) and (3.18) can then be used in the differentiation of (A3.9) and (A3.10) to give

$$\overline{da_k} = v_{l,l} da_k - v_{l,k} da_l \tag{A3.19}$$

and

$$\overline{dv} = v_{k,k} dv \tag{A3.20}$$

Equations (A3.14), (A3.19) and (A3.20) give the material time derivatives of elements of length, area and volume, and can be used to determine the rates of change of physical quantities defined as integrals over material lines, surfaces and volumes. In particular, if the quantity F is expressed as an integral over V of $f dv$, where f is the density of F, then

$$dF/dt = (d/dt) \int_V f dv$$

The integral over the moving region V can be transformed into an integral over the stationary region V_0 by using the rule for change of variables in a multiple integral (Eqn (A2.25), and then the time differentiation can be taken inside the integral sign

$$(d/dt) \int_V f dv = (d/dt) \int_{V_0} f J dV = \int_{V_0} (\dot{f} J + f \dot{J}) dV$$

Using Eqns (A3.10), (A3.17) and (A3.20) it follows that

$$(d/dt) \int_V f dv = \int_V (\dot{f} dv + f \overline{dv}) = \int_V \overline{f dv} = \int_V (\dot{f} + f v_{k,k}) dv$$

In the particular case where $F = M$, the total mass of B, and $f = \rho$, the mass density, then the principle of conversation of mass can be stated as

$$dM/dt = \int_V (\dot{\rho} + \rho v_{k,k}) dv = 0$$

Since this must hold for all parts of B, then

$$\dot{\rho} + \rho v_{k,k} = 0 \tag{A3.21}$$

In a similar fashion, if f_k is a vector field defined over B and F is the *flux* of f_k through a material surface S, so that

$$F = \int_S f_k \, da_k$$

then

$$dF/dt = (d/dt) \int_S f_k \, da_k = \int_S \overline{\dot{f_k \, da_k}} = \int_S (\dot{f}_k + f_k v_{l,l} - f_l v_{k,l}) \, da_k$$

Thus

$$dF/dt = \int_S \overset{*}{f}_k \, da_k \tag{A3.22}$$

where the *convected time derivative* of f_k is defined by

$$\overset{*}{f}_k = \dot{f}_k + f_k v_{l,l} - f_l v_{k,l} \tag{A3.23}$$

or alternatively

$$\overset{*}{f}_k = \partial f/\partial t + v_k f_{l,l} + (v_l f_k)_{,l} - (f_l v_k)_{,l}$$

A3.3 CONSERVATION OF MOMENTUM AND THE STRESS TENSOR

The forces acting in the interior of a material are conventionally divided into short and long range. Typical of the latter is gravitation, which acts throughout the body of the material in such a way that the force on any given element of the body is independent of that on any other element. Consequently the net force on the body due to gravity is simply the sum of all the elemental forces. Short-range forces on the other hand are envisaged as acting at the molecular or atomic level. In a phenomenological approach the precise details of the atomic scale interactions are ignored, and the assumption is made that their effects can be adequately described by the hypothesis of forces, or *surface tractions*, acting across surfaces drawn in the material.

Consider the body B occupying a volume V bounded by a surface S. The external forces acting on the body are in general a combination of long-range or body forces acting throughout V and surface forces or tractions acting over the surface S. (An obvious example is a piece of material at rest on a table or bench, where the body force is that due to gravity and the surface traction is the normal reaction of the table or bench to the weight of the body.) In equilibrium, the sum of the external forces applied must be zero whatever the internal forces in the material, whereas in the dynamical case the sum of the external forces must equal the rate of change of linear momentum of the body

$$(d/dt) \int_V \rho v_k \, dv = \int_V \rho f_k \, dv + \oint_S t_k \, da \tag{A3.24}$$

CONSERVATION OF MOMENTUM AND THE STRESS TENSOR   177

where f_k is the body force per unit mass, t_k are the surface tractions and da is the magnitude of the vector element of area da_k. If n_k is a unit vector normal to da, then d$a_k = n_k$da. Equation (A3.24) expresses the principle of the conversation of linear momentum for the body B.

Now suppose that in equilibrium a surface S_1 be drawn in the body so that the volume V is divided into two portions V_1 and V_2. In general the external forces acting on either portion of the body alone will no longer be balanced, so that it is necessary to assume the existence of forces acting across S_1 to maintain equilibrium. The material in V_1 is taken to exert a force through the surface S_1 on the material in V_2 with an equal and opposite force exerted by the material in V_2 on the material in V_1.

Let da be a small element of area in S_1 with unit normal n_k. Suppose n_k is drawn from V_1 towards V_2. Then the force exerted by the material in V_2 on the material in V_1 through the area da is assumed to be proportional to da and to depend on the orientation n_k of da. Thus the force can be written $t_k(n_l)$da, where the t_k are the components of the *stress vector* in the material. In particular, when da is normal to the coordinate axis Ox_l and the normal n_l lies along the positive Ox_l axis, the components of the stress vector are written t_{kl}.

Now suppose that S_1 is a closed surface drawn entirely within the body and that V_1 denotes the volume enclosed by S_1. At each point of S_1 let the unit normal be drawn pointing out of V_1. The net force acting on the material in V_1 now consists of the integral over V_1 of the body forces plus the integral over S_1 of the stress vector t_k. In equilibrium this force must vanish; in the dynamical case the force must equal the rate of change of momentum of the material in V_1. In either case, if the linear dimensions of V_1 are imagined to decrease without limit, so that V_1 shrinks to a point, it is clear that the volume integrals of momentum and body force will decrease according to the cube of the linear dimensions, whereas the surface integral decreases only as the square. Hence

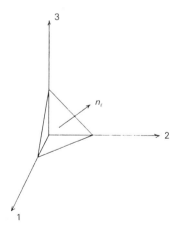

Fig. A3.1 Elementary tetrahedron.

in the limit the surface integral must vanish independently of the volume integrals. For the special case where V_1 is a tetrahedron oriented as in Fig. A3.1, this means that the sum of the forces acting on the faces of the tetrahedron must vanish as the length of the sides tends to zero. If the oblique face has area da and unit normal n_l the face perpendicular to the Ox_l axis has area d$a_l = n_l$da. The forces acting on the tetrahedron across its faces are $t_k(n_l)$da on the oblique face and $-t_{kl}$da_l on the remaining faces, the negative sign arising from the fact that the outward pointing normals on these faces are in the negative direction of the coordinate axes. It follows therefore that

$$t_k(n_l) = t_{kl}n_l \tag{A3.25}$$

Since the product $t_{kl}n_l$d$a = t_{kl}$da_l is the force acting across da, it is a vector quantity. From Section A2.3, da_l is a relative vector of weight $+1$, and it therefore follows that the t_{kl} are the components of a second rank relative tensor of weight -1, known as the *stress tensor*. Knowledge of the stress tensor at all points of a material body is sufficient to allow the determination of the internal forces acting across any arbitrarily oriented element of area within the body.

By the same argument used to demonstrate the existence of the stress tensor, it can easily be shown that on the surface S of B at time t the external tractions t_k must be related to the stress tensor through $t_k = t_{kl}n_l$. Then the divergence theorem (Section A2.4) can be used to transform the surface integral in Eqn (A3.24) to a volume integral, giving

$$(d/dt)\int_V \rho v_k \, dv = \int_V (\rho f_k + t_{kl,l}) \, dv \tag{A3.26}$$

On carrying out the time differentiation and using the conservation of mass as expressed in Eqn (A3.21), this becomes

$$\int_V \rho \dot{v}_k \, dv = \int_V (\rho f_k + t_{kl,l}) \, dv$$

Since this must apply equally to every part of B, it follows that

$$\rho \dot{v}_k = \rho f_k + t_{kl,l} \tag{A3.27}$$

which is the differential form of the conservation of momentum Eqn (A3.24).

A3.4 CONSERVATION OF ANGULAR MOMENTUM

In particle mechanics, the conservation of angular momentum follows from the conservation of linear momentum. This is not the case in continuum mechanics, where the angular momentum principle has to be adopted as an additional law. In the notation of the previous section, it is

$$(d/dt) \int_V \epsilon_{kmp} x_m (\rho v_p) \, dv = \int_V (\epsilon_{kmp} x_m (\rho f_p) + L_k) \, dv +$$
$$\oint_S \epsilon_{kmp} x_m t_p \, da \qquad (A3.28)$$

The left side is the rate of change of the moment of linear momentum about the origin O. The first term in the volume integral on the right is the moment of the body forces about O, and the surface integral represents the moment of the surface tractions. The second term in the volume integral allows for the possibility of *body couples* L_k. The existence of such couples is directly related to the symmetry of the stress tensor.

Substituting $t_p = t_{pq} n_q$ in the surface integral and using the divergence theorem allows the right side of Eqn (A3.28) to be written

$$\int_V [\epsilon_{kmp} x_m (\rho f_p + t_{pq,q}) + (L_k - \epsilon_{kpq} t_{pq})] \, dv$$

The left side reduces to

$$\int_V \epsilon_{kmp} x_m (\rho \dot{v}_p) \, dv$$

Since conservation of linear momentum demands

$$\rho \dot{v}_p = \rho f_p + t_{pq,q}$$

Equation (A3.28) reduces to

$$\int_V (L_k - \epsilon_{kpq} t_{pq}) \, dv = 0$$

and since this must hold for all parts of B, finally

$$L_k - \epsilon_{kpq} t_{pq} = 0 \qquad (A3.29)$$

This is the differential expression of the conservation of angular momentum and implies in particular that, in the absence of body couples L_k, the stress tensor t_{pq} is symmetric.

A3.5 ELECTRODYNAMICS

The motion of a purely elastic material is determined by the conservation laws of mass, momentum and angular momentum and the specific stress–strain relationships that happen to hold for the particular material in question. In a piezoelectric material, the situation is complicated by the interaction of electromagnetic and mechanical phenomena, and the conservation laws have to be complemented by the electromagnetic field equations.

Fortunately, in most situations of practical interest it is permissible to make the *quasistatic* approximation and to neglect the coupling of the electric and magnetic fields that results from the time dependent terms in Maxwell's equations. The justification for this neglect essentially lies in the several orders of magnitude difference between the velocities of elastic waves on the one hand and electromagnetic waves on the other, and amounts to assuming that the electric field is completely described by the static field equations.

In the quasistatic approximation the relevant equations are

$$\oint_S D_k \, da_k = Q \tag{A3.30}$$

$$\oint_C E_k \, dl_k = 0 \tag{A3.31}$$

$$D_k = \epsilon_0 E_k + P_k \tag{A3.32}$$

where S is a closed surface bounding a volume V, and C is a closed curve in V. The quantities D_k, E_k, and P_k are, respectively, the electric displacement, the electric field and the polarization, and Q is the total free charge contained in V. ϵ_0 is the permittivity of free space. The polarization P_k is introduced to represent the electrical properties of any dielectric materials present, and is zero in free space. If the total charge Q is taken to be a scalar quantity, then since the element of area da_k is a relative vector of weight $+1$, it follows from Eqn (A3.30) that D_k must be a relative vector of weight -1. From Eqn (A3.31) and the vector nature of the line element dl_k it similarly follows that E_k is an absolute vector. Additionally from Eqn (A3.32) follows that P_k is a relative vector of weight -1 and ϵ_0 is a relative scalar of weight -1.

If the field quantities are sufficiently smooth in V, the divergence theorem and Stokes' theorem can be applied to Eqns (A3.30) and (A3.31), respectively, to give the differential field equations

$$D_{k,k} = q \tag{A3.33}$$

$$\epsilon_{kmp} E_{m,p} = 0 \tag{A3.34}$$

where it has been assumed that the total charge Q is the integral over V of a free charge density q. It follows from (A3.34) that the field E_k is the gradient of a scalar ϕ known as the electric potential

$$E_k = -\phi_{,k} \tag{A3.35}$$

Equation (A3.33) can be rewritten in terms of E_k and P_k by using Eqn (A3.32)

$$\epsilon_0 E_{k,k} = q - P_{k,k} \tag{A3.36}$$

or in terms of the potential

$$\epsilon_0 \phi_{,kk} = -q + P_{k,k} \tag{A3.37}$$

If the region V contains a surface of discontinuity S_d then the differential field equations cannot be obtained by direct application of the integral theorems. Suppose that S_d divides V into two parts V_1 and V_2, and the unit normal n_k at each point on S_d points from V_1 to V_2. Then application of the divergence theorem to V_1 and V_2 separately leads to the conclusion that Eqn (A3.33) must be satisfied in each region, but that in addition the change in D_k on crossing S_d must satisfy

$$(D_k^+ + D_k^-)n_k = w \tag{A3.38}$$

where w is the surface density of free charge on S_d, and D_k^+, D_k^- are the limiting values of D_k as S_d is approached from within V_2 and within V_1, respectively. Similar arguments applied in Eqn (A3.31) lead to the conclusion that Eqn (A3.34) must be satisfied on either side of S_d, but that in crossing S_d the electric field must additionally satisfy

$$(E_k^+ - E_k^-)m_k = 0 \tag{A3.39}$$

where m_k is any unit vector tangent to S_d. A sufficient condition for this is to require that the potential ϕ be continuous across S_d.

Equation (A3.38) can be rewritten in terms of E_k and P_k

$$\epsilon_0(E_k^+ - E_k^-)n_k = w - (P_k^+ - P_k^-)n_k \tag{A3.40}$$

In the particular case where S_d is the boundary of a dielectric body and P_k^+ is zero, then

$$\epsilon_0(E_k^+ - E_k^-)n_k = w + P_k n_k \tag{A3.41}$$

where P_k is the limiting value of the polarization on the surface S_d. Equations (A3.36) and (A3.41) show clearly the equivalence of the polarized dielectric to a volume charge density $-P_{k,k}$ and a surface charge density $P_k n_k$.

If the charge densities q and w, and the polarization P_k are given, the equations above are sufficient to determine the electric field and displacement E_k and D_k. In a real case, generally the polarization is not given but is rather a quantity to be determined as part of the solution to a given problem. Then in order to solve the field equations it is necessary to assume some functional relationship between P_k and the other field variables. The precise form of the relationship will depend upon the particular problem, specifically on the type of material being considered. In elementary texts, attention is usually restricted to linear isotropic materials and the assumption is made that P_k is proportional to E_k, that is $P_k = kE_k$, where k is a constant. Whatever assumption is made, subsequent discussion is restricted in application to the particular class of materials defined. For present purposes, it is advantageous to delay making any such assumptions and to regard P_k as an independent quantity, except insofar as it is constrained by the field equations themselves.

As in earlier sections of this Appendix, consider then a body B occupying a volume V bounded by a surface S at time t. As before, the points of B are labelled by their coordinates X_K at time $t = t_0$, when B occupied the volume V_0

bounded by the surface S_0. Additionally, let P_k be a polarization field defined in V, let q be the free charge density in V, and let w be the free charge density on S. In terms of the potential ϕ and the field $E_k = -\phi_{,k}$ the field equations are Eqn (A3.37) in V and (A3.41) on S, with ϕ being assumed continuous across S.

$$\epsilon_0 \phi_{,kk} = -q + P_{k,k} \tag{A3.37}$$

$$\epsilon_0(E_k^+ - E_k^-)n_k = w + P_k n_k \tag{A3.41}$$

In terms of the total volume and surface charge densities q_t and w_t defined by $q_t = q - P_{k,k}$ and $w_t = w + P_k n_k$ these become

$$\epsilon_0 \phi_{,kk} = -q_t \tag{A3.42}$$

$$\epsilon_0(E_k^+ - E_k^-)n_k = w_t \tag{A3.43}$$

The physical interpretation of the fields E_k and ϕ is that $E_k(x_l)$ represents the force per unit charge that would be experienced by an infinitesimally small test charge at the position x_l, and that $\phi(x_l)$ represents the potential energy per unit charge. The potential energy can be regarded as equal to the work done per unit charge, against the electric field, in bringing up a test charge to the point x_l from infinity, or, more precisely, from any point where the fields are zero. According to the field-theoretic view, the mutual potential energy of a system of charges is to be regarded as residing in the field itself. Since the field is determined through the field equations by the current distribution of charges, it follows that the precise details of the way in which the charges were assembled are irrelevant to their potential energy.

Consider then a virtual process in which the body B is rigidly clamped so that no physical displacements can occur, and in which the surface and volume charge densities w_t and q_t are incremented by amounts Δq_t and Δw_t. Then the work done in this process, which can be equated to the change ΔU_F in the potential energy of the system, is

$$\Delta U_F = \int_V \phi \Delta q_t \, dv + \oint_S \phi \Delta w_t \, da$$

Writing ΔE_k for the change in the field E_k resulting from the virtual process, then it follows from the field equations Eqns (A3.42) and (A3.43) that

$$\Delta U_F = \int \epsilon_0 E_k \Delta E_k \, dv$$

from which it follows straightforwardly that the total potential energy U_F of the system of charges q_t and w_t is

$$U_F = (\epsilon_0/2) \int E_k^2 \, dv \tag{A3.44}$$

where the integral is understood to be taken over all space.

Although derived by a consideration of a particular virtual process, the result, Eqn (A3.44), is independent of that process. At least in the quasistatic approximation, U_F in Eqn (A3.44) can be regarded as the mutual potential energy, or the electrostatic field energy, even in the case where the dielectric body B is undergoing an arbitrary motion. Then the changes in U_F have to be related to both the mechanical work done by the electrical forces acting on the material and to the changes in *internal energy* of the material resulting from changes in its state of polarization.

The integral in Eqn (A3.44) is to be taken over all space. In calculating the rate of change of U_F it is necessary to take account of the discontinuities in E_k on crossing the surface S of B, and the space integral has to be split into an integral over the volume V of B and an integral over the rest of space V^*. Generally, V and V^* will be time-dependent owing to the motion of B, so it is convenient to be able to transform the integrals to integrals over stationary regions. In the case of V, the equations $x_k = f_k(X_K, t)$ allow this to be done, but in the region V_0^* exterior to V_0 this mapping is not defined. Nevertheless, it is useful to extend the functions f_k arbitrarily over V_0^*, with the sole proviso that they be continuous over the surface S_0.

The rate of change of U_F can then be written

$$dU_F/dt = (d/dt)\int (\epsilon_0/2) E_k^2 \, dv = (\epsilon_0/2) \int \overline{E_k^2 \, dv}$$

Therefore

$$dU_F/dt = \epsilon_0 \int E_k \dot{E}_k \, dv + (\epsilon_0/2) \int E_k^2 v_{l,l} dv \qquad (A3.45)$$

Using Eqn (A3.32) the first term on the right can be written as

$$\epsilon_0 \int E_k \dot{E}_k \, dv = \int E_k \dot{D}_k \, dv - \int E_k \dot{P}_k \, dv \qquad (A3.46)$$

If it is assumed that free charge is conserved in the motion of B, then the total free charge Q in S will be constant. Therefore by Eqn (A3.30) the surface integral of D_k over S will be constant, and by Eqn (A3.22) the convected time derivative $\overset{*}{D}_k$ of D_k will vanish

$$\overset{*}{D}_k = \dot{D}_k + D_k v_{l,l} - D_l v_{k,l} = 0$$

Therefore

$$\int E_k \dot{D}_k \, dv = -\int E_k D_k v_{l,l} dv + \int E_k D_l v_{k,l} dv \qquad (A3.47)$$

The second term in Eqn (A3.46) involves $\dot{P}_k$, the material time derivative of the polarization. This can be rewritten in terms of the polarization per unit mass π_k, with $P_k = \rho \pi_k$. This allows the separation of changes in P_k which are only due to the motion of the material from intrinsic changes in P_k which can be regarded as changes in the state of polarization of the dielectric. In terms of π_k, $\dot{P}_k = \overline{\rho \pi_k} = \dot\rho \pi_k + \rho \dot\pi_k$, and from Eqn (A3.21)

$$\dot{P}_k = -P_k v_{l,l} + \rho \dot\pi_k \qquad (A3.48)$$

The rate of change of the field energy U_F can now be written

$$dU_F/dt = \int \{[E_k D_l - (\epsilon_0/2)E_m^2 \delta_{kl}]v_{k,l} - \rho E_k \dot{\pi}_k\} dv$$

or

$$dU_F/dt = \int \{[t_{kl}^M + E_k P_l]v_{k,l} - \rho E_k \dot{\pi}_k\} dv \tag{A3.49}$$

where the *Maxwell stress tensor* $t_{kl}^M = [\epsilon_0 E_k E_l - (\epsilon_0/2)E_m^2 \delta_{kl}]$ has been introduced. The first two terms on the right of Eqn (A3.49) can be identified with the rate at which mechanical work is done against the electrostatic forces, and the negative of the third term can be identified with the rate at which work is done in changing the polarization state of the material.

A3.6 CONSERVATION OF ENERGY

Consider again a body B in motion under the influence of external body forces f_k per unit mass and surface tractions t_k. Assuming that B consists of a polarized dielectric material carrying volume and surface free charge densities q and w, it will be associated with an electrostatic field E_k having an energy U_F given by Eqn (A3.44). B will also have kinetic energy U_K equal to the integral over V of the kinetic energy density $\rho v_k^2/2$. In addition to the field energy U_F and the kinetic energy U_K it is assumed that B has an internal energy U_I. The internal energy is taken to be the integral over the body of an energy density Σ per unit mass, and it is further assumed that the energy density is a function of the deformation gradients $x_{k,K}$, the polarization per unit mass π_k and the *entropy* per unit mass σ. Then the total energy U of the system consisting of B and its associated electrostatic field is $U = U_I + U_K + U_F$, where

$$U_I = \int_V \rho \Sigma(x_{k,K}, \pi_k, \sigma) dv$$

$$U_K = \int_V [(\rho v_k^2)/2] dv$$

$$U_F = \int_{V+V^*} [(\epsilon_0 E_k^2)/2] dv$$

The principle of *conservation of energy* states that the rate of change of the total system energy U must equal the rate at which external work is done on the system plus the rate at which thermal energy is supplied to the system. The former is just the rate of working of the *external* body forces and surface tractions, and if dissipation is neglected, the latter is the rate at which heat is transmitted through S, that is, the integral over S of the heat flux vector h_k. Therefore

$$dU/dt = \int_V \rho f_k v_k dv + \oint_S t_k v_k da + \oint_S h_k da_k \tag{A3.50}$$

Note that in Eqn (A3.50) the forces f_k and surface tractions t_k are the *external* influences acting, and in particular do not include the electrostatic forces on the material. Hence f_k and t_k do not denote the same quantities as they do in Eqns (A3.24) and (A3.27).

The integral of h_k over S in Eqn (A3.50) can be transformed into a volume integral by using the divergence theorem. Then $(h_{k,k} dv)$ can be interpreted as the rate at which heat or thermal energy is supplied to the volume element dv. In a reversible process in the thermodynamic sense, this can be equated to the product of the absolute temperature θ and the rate of increase of the entropy of the volume element dv, $\overline{\rho\dot{\sigma}dv}$. Since mass is conserved, it follows that

$$\oint_S h_k \, da_k = \int_V \rho\theta\dot{\sigma} \, dv \tag{A3.51}$$

The rate of change of the electrostatic field energy U_F is given by Eqn (A3.49). The rate of change of the kinetic energy U_K is

$$dU_K/dt = (d/dt) \int_V (\rho v_k^2/2) dv = \int_V \rho v_k \dot{v}_k \, dv \tag{A3.52}$$

The rate of change of the internal energy U_I is

$$dU_I/dt = (d/dt) \int_V \rho \Sigma \, dv = \int_V \rho \dot{\Sigma} \, dv$$

But $\dot{\Sigma}$ is

$$\dot{\Sigma} = \frac{\partial \Sigma}{\partial x_{k,K}} \dot{x}_{k,K} + \frac{\partial \Sigma}{\partial \pi_k} \dot{\pi}_k + \frac{\partial \Sigma}{\partial \sigma} \dot{\sigma}$$

and since $\dot{x}_{k,K} = v_{k,l} x_{l,K}$,

$$\rho\dot{\Sigma} = t^L_{kl} v_{k,l} + \rho \frac{\partial \Sigma}{\partial \pi_k} \dot{\pi}_k + \rho \frac{\partial \Sigma}{\partial \sigma} \dot{\sigma}$$

where

$$t^L_{kl} = \rho x_{l,K} \frac{\partial \Sigma}{\partial x_{k,K}} \tag{A3.53}$$

Therefore

$$dU_I/dt = \int_V \left\{ t^L_{kl} v_{k,l} + \rho \frac{\partial \Sigma}{\partial \pi_k} \dot{\pi}_k + \rho \frac{\partial \Sigma}{\partial \sigma} \dot{\sigma} \right\} dv \tag{A3.54}$$

Transforming the first term by using the divergence theorem gives

$$dU_I/dt = \int_V \left\{ -t^L_{kl,l} v_k + \rho \frac{\partial \Sigma}{\partial \pi_k} \dot{\pi}_k + \rho \frac{\partial \Sigma}{\partial \sigma} \dot{\sigma} \right\} dv +$$

$$\oint_S t^L_{kl} v_k n_l \, da \tag{A3.55}$$

The first term of Eqn (A3.49) can be similarly transformed, with the proviso that since the integral in Eqn (A3.49) extends over all space, the resulting surface integral involves the 'jump' in $t_{kl}^M + E_k P_l$ in crossing S

$$dU_F/dt = \int_{V+V^*} \{-[t_{kl,l}^M + (E_k P_l)_{,l}]v_k - \rho E_k \dot{\pi}_k\} dv -$$

$$\oint_S [\![t_{kl}^M + E_k P_l]\!] v_k n_l \, da \tag{A3.56}$$

where the notation $[\![f]\!]$ is used to indicate the 'jump' in a quantity f in crossing a surface of discontinuity.

Equations (A3.51), (A3.52), (A3.55) and (A3.56) can now be substituted into Eqn (A3.50) to give an equation involving the sum of three volume integrals and a surface integral

$$\int_{V+V^*} [\rho \dot{v}_k - \rho f_k - t_{kl,l}^L - t_{kl,l}^M - (E_k P_l)_{,l}] v_k \, dv +$$

$$\oint_S \{-t_k + t_{kl}^L n_l - [\![t_{kl}^M n_l + E_k P_l n_l]\!]\} v_k \, da +$$

$$\int_V \left[\rho \frac{\partial \Sigma}{\partial \pi_k} - \rho E_k \right] \dot{\pi}_k \, dv +$$

$$\int_V \left[\rho \frac{\partial \Sigma}{\partial \sigma} - \rho \theta \right] \dot{\sigma} \, dv$$

This sum must vanish for all admissible velocity, polarization and entropy fields and therefore the terms multiplying $v_k, \dot{\pi}_k$ and $\dot{\sigma}$ can be separately set equal to zero, leading to the field equations

$$\rho \dot{v}_k = \rho f_k + t_{kl,l}^L + t_{kl,l}^M + (E_k P_l)_{,l} \tag{A3.57}$$

$$\frac{\partial \Sigma}{\partial \pi_k} = E_k \tag{A3.58}$$

$$\frac{\partial \Sigma}{\partial \sigma} = \theta \tag{A3.59}$$

satisfied in V,

$$t_{kl,l}^M = 0 \tag{A3.60}$$

satisfied in V^*, and

$$t_k = t_{kl}^L n_l - [\![t_{kl}^M + E_k P_l]\!] n_l \tag{A3.61}$$

satisfied on S.

Comparing Eqn (A3.57) to Eqn (A3.27) leads naturally to the physical interpretation of t_{kl}^L as the local 'elastic' stress in the material, related to the deformation by Eqn (A3.53), and to the interpretation of $t_{kl}^M + E_k P_l$ as the electrostatic stress. From the definition of the Maxwell stress tensor t_{kl}^M and the field Eqns (A3.34) and (A3.36) it follows that

$$t_{kl,l}^M + (E_k P_l)_{,l} = qE_k + P_l E_{k,l}$$

The first term is just the force exerted by the field E_k on the free charge density q, and the second is the force on an electric dipole P_l in an electric field E_k.

It should be noted that for a dipole in an electric field, not only is there a force $P_l E_{k,l}$ acting, but there is also a couple $\epsilon_{klm} P_l E_m$. This couple constitutes a body couple in the sense of Section A3.4, and its existence implies that the local stress t_{kl}^L must be asymmetric.

This can be confirmed by considering the dependence of the internal energy density Σ on its arguments $x_{k,K}$, π_k, and σ. Σ is an absolute scalar and therefore invariant under orthogonal transformations of coordinates. However, the deformation gradients and the polarization are vector quantities and transform according to the appropriate law under orthogonal transformation. It is therefore clear that Σ cannot be an arbitrary function of its arguments. It can in fact be shown (Weyl, 1939) that any scalar function of N vectors must be expressible as a function of the scalar products of the N vectors taken two at a time, so that in the present case Σ must be expressible as a function of the quantities $x_{k,K} x_{k,L}$, $x_{k,K} \pi_k$, and $\pi_k \pi_K$. The quantities $x_{k,K} x_{k,L}$ are just the components of the deformation tensor C_{KL}, and if P_K is written for $x_{k,K} \pi_k$ and π^2 for $\pi_k \pi_k$, then Σ must be expressible as a function of C_{KL}, P_K, and π^2. The defining relation for P_K can be inverted to give $\pi_k = P_K X_{K,k}$, from which it follows that

$$\pi^2 = X_{K,k} X_{L,k} P_K P_L = C_{KL}^{-1} P_K P_L$$

where $C_{KL}^{-1} = X_{K,k} X_{L,k}$ is the inverse of the deformation tensor. Thus π^2 is itself a function of the C_{KL} and the P_K and so can be dropped from the argument list of Σ, leading to the conclusion that Σ must be expressible as a function of the C_{KL} and P_K (in addition to the scalar quantity σ). The partial derivatives of Σ with respect to $x_{k,K}$ and π_k can then be expressed as

$$\frac{\partial \Sigma}{\partial x_{k,K}} = \frac{\partial \Sigma}{\partial C_{LM}} \frac{\partial C_{LM}}{\partial x_{k,K}} + \frac{\partial \Sigma}{\partial P_L} \frac{\partial P_L}{\partial x_{k,K}}$$

$$\frac{\partial \Sigma}{\partial \pi_k} = \frac{\partial \Sigma}{\partial P_L} \frac{\partial P_L}{\partial \pi_k}$$

Then from Eqns (A3.53) and (A3.58) it follows that

$$t_{kl}^L = \rho(x_{l,M} x_{k,L} + x_{k,M} x_{l,L}) \frac{\partial \Sigma}{\partial C_{LM}} + P_k E_l$$

so that the antisymmetric part $t_{[kl]}^L$ is just

$$t_{[kl]}^L = P_{[k} E_{l]}$$

just as required by the conservation of angular momentum Eqn (A3.29).

A3.7 SUMMARY OF EQUATIONS FOR THE ELASTIC DIELECTRIC

The results of the previous sections can be summarized as follows. For an elastic, heat-conducting, dielectric body B, without dissipation, occupying a volume V bounded by a surface S and acted upon by external forces f_k per unit mass and surface tractions t_k per unit area, the field equations in V are

$$D_{k,k} = q \qquad (A3.33)$$
$$\rho \dot{v}_k = \rho f_k + t^L_{kl,l} + (t^M_{kl} + E_k P_l)_{,l} \qquad (A3.57)$$

while on the boundary surface S

$$[\![D_k n_k]\!] = w \qquad (A3.38)$$
$$[\![\phi]\!] = 0 \qquad (A3.39)$$
$$t_k = t^L_{kl} n_l - [\![t^M_{kl} + E_k P_l]\!] n_l \qquad (A3.61)$$

where $[\![f]\!]$ denotes the 'jump' in a quantity f in crossing S. In the region V^* exterior to V,

$$D_{k,k} = 0 \qquad (A3.62)$$

In the above q and w are volume and surface free charge densities in V and on S. E_k, D_k and P_k are respectively the electric field, displacement and polarization, with

$$D_k = \epsilon_0 E_k + P_k \qquad (A3.32)$$

where ϵ_0 is the permittivity of free space. P_k is identically zero in V^*. The electric field E_k is obtained from the potential ϕ via

$$E_k = -\phi_{,k} \qquad (A3.35)$$

The Maxwell stress tensor is defined in terms of the electric field by

$$t^M_{kl} = \epsilon_0 E_k E_l - (\epsilon_0/2) E_m^2 \delta_{kl} \qquad (A3.63)$$

The mass density of B is ρ, and v_k is the velocity of a point of B. The body B is assumed to have an internal energy density Σ per unit mass, given as a function of the deformation gradients $x_{k,K}$, the polarization per unit mass $\pi_k = P_k/\rho$, and the entropy per unit mass σ. Then the *constitutive relations* for the local elastic stress t^L_{kl}, the electric field E_k and the absolute temperature which follow from the conservation of energy are

$$t^L_{kl} = \rho x_{l,K} \frac{\partial \Sigma}{\partial x_{k,K}} \qquad (A3.53)$$

$$E_k = \frac{\partial \Sigma}{\partial \pi_k} \qquad (A3.58)$$

SUMMARY OF EQUATIONS FOR THE ELASTIC DIELECTRIC 189

$$\theta = \frac{\partial \Sigma}{\partial \sigma} \tag{A3.59}$$

Given the form of the internal energy function Σ, these equations are sufficient to determine the motion of B subject to the external influences f_k and t_k and the surface and volume charge densities q, w, which are assumed to be conserved in the motion.

Appendix 4
Linear piezoelectric theory

A4.1 LINEAR FIELD EQUATIONS

The field equations and boundary conditions for an elastic dielectric set out in Appendix 3 are, subject to certain assumptions, valid for arbitrary electric fields and finite deformations. The assumptions were

(a) that the electromagnetic field could be adequately described by the quasi-static electric field equations;
(b) that the material properties of the dielectric could be described in terms of an internal energy function; and
(c) that dissipation of energy could be neglected.

The equations are therefore sufficiently general to be used to describe electrostrictive phenomena and other non-linear effects. What may be termed the classical theory of piezoelectricity is based on a restricted set of equations, derived from those of Appendix 3 by making a general assumption of linearity.

More specifically, the assumption of linearity amounts to the neglect of any terms in the field equations that are non-linear in either the fields themselves or their spatial and temporal derivatives. An immediate consequence of this sweeping assumption is that the electrostatic stress $t_{kl}^M + E_k P_l$, which involves the product of the electric field quantities, is discarded, thus making it impossible for the linear theory to include electrostrictive effects. A further consequence is that the material time derivative of a field quantity f reduces to the partial time derivative, and that the distinction between the initial and current coordinates of a material particle X is lost, so that the quantities $f_{,K}$ and $f_{,k}$ can no longer be distinguished. Thus in the linear theory it is unnecessary to maintain the notational conventions of using upper and lower case indices to separate quantities referred to initial and current coordinates of a point X respectively.

These assumptions can be made more precise by introducing the particle displacement $u_k(X_K)$ such that by definition

$$u_k(X_K) = x_k(X_K) - \delta_{kK} X_K$$

LINEAR FIELD EQUATIONS

Then both u_k and its derivatives are assumed to be sufficiently small that only linear terms need be retained. From the definition

$$x_{k,K} = \delta_{kK} + u_{k,K}$$

so that for any field quantity f

$$f_{,K} = f_{,k} x_{k,K} = f_{,k}\delta_{kK} + f_{,k}u_{k,K} \sim f_{,k}\delta_{kK}$$

and

$$\dot{f} = \partial f/\partial t + v_k f_{,k} \sim \partial f/\partial t$$

Also from the definition of the Jacobian determinant J as $\det\{x_{k,K}\}$ it follows that up to linear terms in the displacement gradients $u_{k,K}$

$$J = \det\{\delta_{kK} + u_{k,K}\} \sim 1 + \delta_{kK} u_{k,K}$$

The principle of conservation of mass (Eqn (A3.21)) can be written $\overline{(\rho J)} = 0$, or in terms of the mass density ρ_0 in the initial state, $\rho J = \rho_0$. Substituting for J yields to first order in the displacement gradients

$$\rho = \rho_0 (1 + \delta_{kK} u_{k,K})$$

The mechanical field equations (A3.57) and (A3.61) can now be written in the linear approximation

$$\rho_0 \dot{v}_k = \rho f_k - t^L_{kl,l} \tag{A4.1}$$

and

$$t_k = t^L_{kl} n_l \tag{A4.2}$$

where now the superimposed dot is used to denote the ordinary partial time derivative rather than the material time derivative.

The stress tensor t^L_{kl} is derived from the internal energy density Σ by Eqn (A3.53). As shown in Section A3.6, Σ can be regarded as a function of the quantities C_{KL}, P_K and the entropy per unit mass σ, where

$$C_{KL} = x_{k,K} x_{k,L} \qquad P_K = \pi_k x_{k,K}$$

and π_k is the polarization per unit mass. Introducing the displacement gradients and retaining linear terms only,

$$C_{KL} = (\delta_{kK} + u_{k,K})(\delta_{kL} + u_{k,L}) \sim \delta_{KL} + \delta_{kK} u_{k,L} + u_{k,K}\delta_{kL}$$
$$P_K = (\delta_{kK} + u_{k,K})\pi_k \sim \delta_{kK}\pi_k$$

In this approximation the Lagrangian finite strain tensor reduces to the *infinitesimal strain tensor* of classical elasticity theory

$$S_{KL} = (C_{KL} - \delta_{KL})/2 \sim (\delta_{kK} u_{k,L} + u_{k,K}\delta_{kL})/2$$

and Σ can then be written as a function of S_{KL}, P_K and σ. Equation (A3.53) can then be replaced by

$$t_{kl}^L = \rho x_{l,k} \frac{\partial \Sigma}{\partial x_{k,K}} \sim \rho \delta_{l,K} \frac{\partial \Sigma}{\partial S_{LM}} \frac{\partial S_{LM}}{\partial u_{k,K}} = \rho \frac{\partial \Sigma}{\partial S_{KL}} \delta_{kK} \delta_{iL}$$

Writing $U = \rho \Sigma$ for the internal energy density per unit *volume*

$$t_{kl}^L = \frac{\partial U}{\partial S_{KL}} \delta_{kK} \delta_{iL}$$

If U is regarded as a function of the strains S_{KL} and the polarization P_k and entropy S per unit *volume* rather than the polarization π_k and entropy σ per unit mass, then Eqns (A3.58) and (A3.59) can be written

$$E_k = \frac{\partial \Sigma}{\partial \pi_k} = \frac{\partial U}{\partial P_k}$$

$$T = \frac{\partial \Sigma}{\partial \sigma} = \frac{\partial U}{\partial S}$$

where T has been written for the absolute temperature rather than θ.

Dropping the distinction between initial and current coordinates, the linear field equations can be summarized as follows. For a body B occupying a volume V bounded by a surface S, the electrostatic field equations are

$$D_{k,k} = q \qquad E_k = -\phi_{,k} \qquad D_k = \epsilon_0 E_k + P_k \tag{A4.3}$$

in V and

$$[\![D_k n_k]\!] = w \qquad [\![\phi]\!] = 0 \tag{A4.4}$$

on S, where E_k, D_k, P_k and ϕ are the electric field, displacement, polarization and potential, and q and w are volume and surface densities of free charge. In the space outside V

$$D_{k,k} = 0 \qquad E_k = -\phi_{,k} \qquad D_k = \epsilon_0 E_k \tag{A4.5}$$

The mechanical field equations are

$$\rho_0 \dot{v}_k = \rho f_k - t_{kl,l} \tag{A4.6}$$

in V and

$$t_k = t_{kl} n_l \tag{A4.7}$$

on S, where f_k are the external body forces per unit mass and t_k are the surface tractions acting across S. The stress t_{kl}, where the suffix L has been dropped, is obtained from an internal energy density U given as a function of the strain tensor S_{kl}, the polarization P_k and the entropy S via

$$t_{kl} = \partial U / \partial S_{kl} \tag{A4.8}$$

with $S_{kl} = (u_{k,l} + u_{l,k})/2$. The electric field and polarization are related by

$$E_k = \partial U / \partial P_k \tag{A4.9}$$

and the absolute temperature T is given by

$$T = \partial U/\partial S \qquad (A4.10)$$

A4.2 LINEAR CONSTITUTIVE RELATIONS

The field equations given in the previous section cannot be solved unless a particular functional form is adopted for the internal energy U. The assumption that U is a function of the *deformation* of the material implies a choice of a reference which can be identified with an *undeformed* state. Of course from a purely formal point of view the choice of reference state is entirely arbitrary, but on physical grounds it is natural to require that the reference state corresponds to one of *zero stresses*, mechanical and electrical. The assumption that such a state exists is far from trivial, and amounts to excluding from consideration materials such as *ferroelectrics*, which are permanently polarized, or materials which have stresses and strains 'built-in' as a result of their previous history.

The classical theory of piezoelectricity assumes that at any given temperature T_0, a reference state exists for which the strain S_{kl}, the polarization P_k, the stress t_{kl} and the electric field E_k are all simultaneously zero. If U_0 and S_0 are respectively the values of the internal energy and the entropy in the reference state, then for small, reversible changes it is assumed that U can be adequately expressed as a Taylor series up to and including second order terms in the strains and polarization. (For applications to problems in waves and vibrations, it is usual to assume that the changes are taking place sufficiently rapidly for thermal conduction to be ignored, that is, that they are *adiabatic* or *isentropic*. The entropy S is thus regarded as constant.) Therefore

$$U - U_0 = U^0_{kl}S_{kl} + U^0_k P_k + \tfrac{1}{2}U^0_{kl,mn}S_{kl}S_{mn} + U^0_{kl,m}S_{kl}P_m + \tfrac{1}{2}U^0_{k,l}P_k P_l$$

where

$$U^0_{kl} = \partial U/\partial S_{kl} \qquad U^0_k = \partial U/\partial P_k$$
$$U^0_{kl,mn} = \partial^2 U/\partial S_{kl}\partial S_{mn} \qquad U^0_{kl,m} = \partial^2 U/\partial S_{kl}\partial P_m$$
$$U^0_{k,l} = \partial^2 U/\partial P_k \partial P_l$$

and all the partial derivatives are evaluated at the reference state.

From Eqns (A4.8) and (A4.9) and the assumption that both the stress and field are zero in the reference state, it follows that the first order derivatives U^0_{kl} and U^0_k vanish, so that therefore

$$U - U_0 = \tfrac{1}{2}U^0_{kl,mn}S_{kl}S_{mn} + U^0_{kl,m}S_{kl}P_m + \tfrac{1}{2}U^0_{k,l}P_k P_l \qquad (A4.11)$$

Since all the derivatives are evaluated at the reference state, they can be regarded as constants in the consideration of small changes from the reference. In order to distinguish these *material constants* from others to be introduced later, the following definitions are introduced:

$$c^P_{klmn} = U^0_{kl,mn}$$
$$a_{mkl} = -U^0_{kl,m} \tag{A4.12}$$
$$\chi^{-1}_{kl} = U^0_{k,l}$$

Then U can be written

$$U - U_0 = \tfrac{1}{2}c^P_{klmn}S_{kl}S_{mn} - a_{klm}S_{kl}P_m + \tfrac{1}{2}\chi^{-1}_{kl}P_kP_l \tag{A4.13}$$

Equations (A4.8) and (A4.9) now become

$$t_{kl} = c^P_{klmn} S_{mn} - a_{mkl} P_m$$
$$E_k = -a_{klm} S_{lm} + \chi^{-1}_{kl} P_l \tag{A4.14}$$

These are the fundamental linear constitutive relations for a piezoelectric material. The c^P_{klmn} are the *elastic constants* at constant polarization, the a_{mkl} are *piezoelectric constants* and the χ^{-1}_{kl} are the components of the *inverse susceptibility tensor* at constant strain. Under rotations of the coordinate system, the cs, as, and χ^{-1}'s transform as fourth, third and second rank tensors respectively.

Because of the form of the field equations Eqns (A4.3) to (A4.7) it is advantageous to recast the constitutive relations into a form where the independent variables are the strain and the electric field, and the dependent variables are the stress and the electric displacement. This is most easily done by forming the *electric enthalpy* $H = U - E_k P_k$. Then the total differential of H is

$$dH = dU - E_k\, dP_k - P_k\, dE_k$$
$$= (\partial U/\partial S_{kl})dS_{kl} + (\partial U/\partial P_k)dP_k - E_k\, dP_k - P_k\, dE_k$$

But from Eqn (A4.9), $E_k = (\partial U/\partial P_k)$ so

$$dH = (\partial U/\partial S_{kl})dS_{kl} - P_k\, dE_k$$

Therefore H can be regarded as a function of S_{kl} and E_k with

$$(\partial H/\partial S_{kl}) = (\partial U/\partial S_{kl}) = t_{kl}$$
$$(\partial H/\partial E_k) = -P_k \tag{A4.15}$$

Just as in the case of the internal energy U, H is then expanded in a Taylor series, but now with independent variables S_{kl} and E_k. The linear terms drop out because of Eqns (A4.15) and the assumption of zero stress and field in the reference state, leaving

$$H - H_0 = \tfrac{1}{2}c^E_{klmn}S_{kl}S_{mn} - e_{klm}E_kS_{lm} - \tfrac{1}{2}\chi_{kl}E_kE_l \tag{A4.16}$$

which is analogous to Eqn (A4.13). The material constants c^E_{klmn} are the elastic constants at constant field, the e_{klm} are piezoelectric constants, and the χ_{kl} are the components of the susceptibility tensor. The constants are defined in an obvious manner in terms of the second derivatives of H evaluated at the reference state. Using Eqn (A4.16) in Eqn (A4.15) leads to the constitutive relations

$$t_{kl} = c^E_{klmn} S_{mn} - e_{jkl} E_j \qquad (A4.17)$$

$$P_k = e_{klm} S_{lm} + \chi_{kl} E_l \qquad (A4.18)$$

Adding a term $\epsilon_0 E_k$ to both sides of the second equation, Eqn (A4.18), leads to

$$D_k = e_{klm} S_{lm} + \epsilon^S_{kl} E_l \qquad (A4.19)$$

where ϵ^S_{kl} are the dielectric constants at constant strain, or the 'clamped' dielectric constants.

Equations (A4.17) and (A4.19) are the constitutive relations in their most commonly used form. Other sets useful in special circumstances can be arrived at by defining thermodynamic functions other than the enthalpy and going through the same procedure of expanding in a Taylor series. In particular, if the function $G = U - E_k P_k - t_{kl} S_{kl}$ is used, the resulting constitutive relations take the form

$$S_{kl} = s^E_{klmn} t_{mn} + d_{jkl} E_j$$
$$D_k = d_{klm} t_{lm} + \epsilon^t_{kl} E_l$$

where the s^E_{klmn} and ϵ^t_{kl} are the *elastic compliances* at constant field and the dielectric constants at constant stress (or 'free' dielectric constants) respectively, and the d_{klm} are a further set of piezoelectric constants.

Because of their definitions in terms of the second derivatives of thermodynamic functions of state, and the rule that the order of partial differentiation can be changed provided that the functions involved are sufficiently smooth, the material constants have the following symmetries. First the dielectric constants are symmetric, $\epsilon_{kl} = \epsilon_{lk}$. Second, for both elastic constants and compliances, $c_{klmn} = c_{mnkl}$. In addition, since both stress and strain tensors are symmetrical, the elastic constants have the additional symmetries $c_{klmn} = c_{lkmn}$, similar relations holding for the compliances. For the same reason, the piezoelectric contants are symmetric in their second and third indices, for example, $e_{klm} = e_{kml}$.

A4.3 POSITIVE-DEFINITE CHARACTER OF THE INTERNAL ENERGY U

Since the reference state for the material has been assumed to be one such that all fields and stresses, both electrical and mechanical are zero, it is natural to assume that any changes from the reference state will require a definite amount of external work and a consequent increase in the internal energy U. Therefore in Eqn (A4.13), $U - U_0$ must be positive-definite.

U can be expressed as a function of the strain S_{kl} and the field E_k by using $H(S_{kl}, E_k) = U - E_k P_k$ and substituting for P_k from Eqn (A4.18). The result is

$$U - U_0 = \tfrac{1}{2} c_{klmn} S_{kl} S_{mn} + \tfrac{1}{2} \chi_{kl} E_k E_l$$

In particular, for $E_k = 0$ it follows that $\tfrac{1}{2} c_{klmn} S_{kl} S_{mn}$ is separately positive-definite, that is, is positive for all non-zero S_{kl}. If S_{kl} is replaced by the product $x_k y_l$ then the positive-definite character is still retained, and if y_k is then replaced by $y_k = y n_k$ where n_k is a unit vector and y an arbitrary scalar, then $U - U_0$ formally reduces to

$$U - U_0 = \tfrac{1}{2} (c_{klmn} n_l n_n x_k x_m) y^2$$

Defining $L_{km} = c_{klmn} n_l n_n$, and noting that $y^2 > 0$, it then follows that $L_{km} x_k x_m$ is also positive definite.

This is an important result, since as shown in Chapter 2, the tensor L_{km} appears in the analysis of plane wave propagation in anisotropic materials, and its positive-definite character is essential to the existence of real phase velocities.

Appendix 5
Coordinate transformations and crystal symmetries

A5.1 ROTATIONS OF THE COORDINATE SYSTEM

In Appendix 2 it was shown that the general form of the transformation relating the coordinates x_k, x'_k of a point P in two rectangular cartesian coordinate systems Ox_k and Ox_k, with a common origin O is

$$x_k = a_{km} x'_m \tag{A5.1}$$

where

$$a_{km} a_{kn} = \delta_{mn} \tag{A5.2}$$

and the inverse of Eqn (A5.1) is

$$x'_k = a_{mk} x_m \tag{A5.3}$$

It was also stated in Appendix 2 that these equations describe either a rotation of the coordinate system, or a rotation accompanied by an inversion. In the first case, the determinant $\det\{a_{km}\} = +1$, in the second $\det\{a_{km}\} = -1$. In the rest of this Appendix, only rotations are considered, so always $\det\{a_{km}\} = +1$.

An alternative description of a rotation of the coordinate system can be given by specifying the axis about which the rotation is to take place along with the magnitude of the rotation. Clearly, for any point on the axis of the rotation, the coordinates x_k and x'_k will be identical, so that from Eqns (A5.1) and (A5.4)

$$x_k = a_{km} x_m, \qquad x_k = a_{mk} x_m \tag{A5.4}$$

In particular, for a rotation about the original coordinate axis Ox_1, the unit vector with components (1,0,0) is invariant, leading to the special form for the matrix A with elements a_{km}

$$\begin{pmatrix} 1 & 0 & 0 \\ 0 & a_{22} & a_{23} \\ 0 & a_{32} & a_{33} \end{pmatrix}$$

Using the orthogonality relations Eqn (A5.2) and the condition on the determinant of A, the matrix further reduces to

$$\begin{pmatrix} 1 & 0 & 0 \\ 0 & c & -s \\ 0 & s & c \end{pmatrix} \tag{A5.5}$$

where $c = \cos(\theta)$ and $s = \sin(\theta)$ for an angle θ which corresponds to the amount by which the original Ox_2 axis is rotated towards Ox_3, as illustrated in Fig. A5.1.

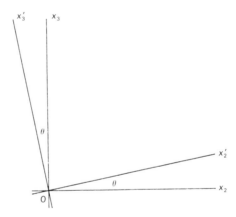

Fig. A5.1 Rotation about Ox_1.

The corresponding matrices for rotations about the other two axes are

$$\begin{pmatrix} c & 0 & -s \\ 0 & 1 & 0 \\ s & 0 & c \end{pmatrix} \tag{A5.6}$$

for rotation about Ox_2 and

$$\begin{pmatrix} c & -s & 0 \\ s & c & 0 \\ 0 & 0 & 1 \end{pmatrix} \tag{A5.7}$$

for rotation about Ox_3.

Rotations about some axis other than one of the coordinate axes are

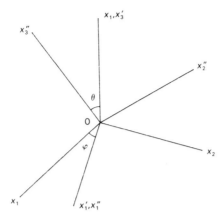

Fig. A5.2 Coordinate system for a doubly rotated plate.

usually regarded as composite transformations, built up of a succession of simple rotations. In the present context it is sufficient to consider two successive rotations, as used to describe the *double rotated* crystal plates. As shown in Fig. A5.2, the coordinate system Ox_k'' associated with a double rotated plate can be thought of as being obtained from the crystallographic system by a composition of, first, a rotation ϕ about the Z or Ox_3 axis to form an intermediate system Ox_k', followed by a second rotation θ about the Ox_1' axis. If x_k, x_k' and x_k'' are the coordinates of a point in the three systems, and if a'_{km} and a''_{km} are the first and second transformations, then

$$x_k = a'_{km} x'_m$$
$$x'_m = a''_{mn} x''_n$$

so that combining the two

$$x_k = a'_{km} a''_{mn} x''_n = a^*_{kn} x''_n \tag{A5.8}$$

Thus the matrix of the composite rotation, say A^*, is just the matrix product of the matrices A' and A'' of the component transformations

$$\begin{pmatrix} c_\phi & -s_\phi & 0 \\ s_\phi & c_\phi & 0 \\ 0 & 0 & 1 \end{pmatrix} \begin{pmatrix} 1 & 0 & 0 \\ 0 & c_\theta & -s_\theta \\ 0 & s_\theta & c_\theta \end{pmatrix}$$

with the obvious notation $c_\phi = \cos(\phi)$ etc. Multiplying out the matrices gives A^* as

$$\begin{pmatrix} \cos(\phi) & -\sin(\phi)\cos(\theta) & \sin(\phi)\sin(\theta) \\ \sin(\phi) & \cos(\phi)\cos(\theta) & -\cos(\phi)\sin(\theta) \\ 0 & \sin(\theta) & \cos(\theta) \end{pmatrix}$$

In the plate coordinate system the plate normal has components δ_{2k} and therefore the components of the plate normal in the original system of the crystal axes are $n_k = a^*_{km}\delta_{2m} = a^*_{k2}$. Thus

$$n_1 = -\sin(\phi)\cos(\theta)$$
$$n_2 = \cos(\phi)\cos(\theta)$$
$$n_3 = \sin(\theta) \quad \quad (A5.9)$$

For the rotated Y-cuts, the angle $\phi = 0$ and then n_k reduces to

$$n_1 = 0$$
$$n_2 = \cos(\theta)$$
$$n_3 = \sin(\theta) \quad \quad (A5.10)$$

A5.2 CRYSTAL SYMMETRIES

The optic or Z axis of quartz is a trigonal or threefold axis. The electric or X axis is a digonal or twofold axis. Because of the presence of the trigonal axis perpendicular to the X axis, the latter is repeated in the plane normal to Z at intervals of 120°, so consequently there are three electric axes, all equivalent. The trigonal axis is not repeated by the twofold axis, since a 180° rotation about any of the X axes transforms the trigonal axis into itself. The twofold symmetry does however ensure that the trigonal axis is not polar. The single trigonal and three digonal axes completely describe the symmetry of quartz.

The presence of these symmetry elements means that the material properties of quartz, as described by the arrays of dielectric, piezoelectric and elastic constants introduced in Appendix 4, must satisfy certain conditions. Consider, for example a rotation of the crystal through 120° about the optic axis. This is a symmetry operation and hence the material constants for the rotated material should be identical to those of the unrotated material. But a rotation of the material relative to a fixed coordinate system is exactly equivalent to an opposite rotation of the coordinate system keeping the crystal fixed, and then the consequence of symmetry is that the material constants in the rotated coordinate system should be identical to those in the original system. If the rotation of the coordinate system is represented by Eqn (A5.1), and if a particular material property is represented by a tensor with components $t_{kmp..}$, then using the tensor transformation law given in Section A2.2, symmetry requires

$$t_{kmp..} = a_{kl}a_{mn}a_{pq}\cdot t_{lnq..} \quad \quad (A5.11)$$

A5.3 SECOND-RANK TENSOR PROPERTIES

The dielectric constants and the thermal expansion coefficients are examples of material properties represented by second-rank tensors. The symmetry conditions for these quantities take the form

$$t_{km} = a_{kl}a_{mn}t_{ln} \tag{A5.12}$$

Suppose first that the a_{km} correspond to a 180° rotation about the X or Ox_1 axis. From Eqn (5.5) the corresponding matrix A_X is

$$\begin{pmatrix} 1 & 0 & 0 \\ 0 & 1 & 0 \\ 0 & 0 & -1 \end{pmatrix} \tag{A5.13}$$

Writing Eqn (A5.12) in matrix form, with T the matrix with elements t_{km} and $\tilde{A}$ denoting the transpose of A, gives

$$T = AT\tilde{A} \tag{A5.14}$$

Carrying out the multiplications leads to the matrix identity

$$\begin{pmatrix} t_{11} & t_{12} & t_{13} \\ t_{21} & t_{22} & t_{23} \\ t_{31} & t_{32} & t_{33} \end{pmatrix} = \begin{pmatrix} t_{11} & -t_{12} & -t_{13} \\ -t_{21} & t_{22} & t_{23} \\ -t_{31} & t_{32} & t_{33} \end{pmatrix}$$

and thus T reduces to

$$\begin{pmatrix} t_{11} & 0 & 0 \\ 0 & t_{22} & t_{23} \\ 0 & t_{32} & t_{33} \end{pmatrix} \tag{A5.15}$$

If now the symmetry operation of a 120° rotation about Z or Ox_3, with matrix A given by Eqn (A5.7), is considered, Eqn (A5.14) leads to

$$T = \begin{pmatrix} t_{11}c^2 + t_{22}s^2 & (t_{11} - t_{22})cs & -st_{23} \\ (t_{11} - t_{22})cs & t_{11}s^2 + t_{22}c^2 & ct_{23} \\ -st_{32} & ct_{32} & t_{33} \end{pmatrix}$$

Since $c = \cos(120)$ and $s = \sin(120)$ are both non-zero it follows that finally T has the form

$$\begin{pmatrix} t_{11} & 0 & 0 \\ 0 & t_{11} & 0 \\ 0 & 0 & t_{33} \end{pmatrix} \tag{A5.16}$$

Hence any second-rank tensor property in a material of symmetry type 32 is, when referred to the crystallographic axes, represented by a diagonal matrix with just two independent components.

A5.4 THIRD- AND FOURTH-RANK TENSOR PROPERTIES: MATRIX NOTATION

The application of the symmetry conditions to second-rank tensors is straightforward, but the situation is complicated with higher-rank tensors because of the large number of components and the fact that matrix methods cannot be applied directly. Both difficulties can be removed by making use of the symmetries of the higher-rank tensors representing the piezoelectric and elastic properties of a material to reduce the number of indices. These symmetries were discussed in Section A4.2. Taking the elastic constants c_{klmn} as an example, the symmetry of the stress and strain tensors results in symmetry in the pairs of indices kl and mn, while the definition of the cs as second derivatives of a function of state results in symmetry between the pairs. Thus $c_{klmn} = c_{lkmn} = c_{mnkl}$. These symmetries reduce the number of independent cs from $3^4 = 81$ to 21.

The reduced notation consists in replacing pairs of indices km, with k, m running from 1 to 3, with single indices $K, M \ldots$ running from 1 to 6 according to the following rules

K:	1	2	3	4	5	6
km:	11	22	33	23 or 32	31 or 13	12 or 21

The rule is applied directly to the stress t_{kl}, the elastic constants c_{klmn} and the piezoelectric constants e_{klm}, to give the reduced stresses t_K, elastic constants c_{KL} and piezoelectric constants e_{kL}. (Note only the second and third indices of the e_{klm} are reduced.) In their reduced form, the stresses can be represented by a 6×1 column matrix T, the elastic constants by a 6×6 square matrix c and the piezoelectric constants by a 3×6 rectangular matrix e.

For the strains S_{kl}, the elastic compliances s_{klmn} and the piezoelectric constants d_{klm}, the rule is modified by introducing factors of 2 when the index pair to be replaced has unequal elements. Thus for the strains, $S_K = S_{kl}$ when $k = l$ but otherwise $S_K = 2S_{kl}$. For the compliances, $s_{KM} = s_{klmn}$ when both $k = l$ and $m = n$; if only one of these conditions hold, then $s_{KM} = 2s_{klmn}$, while if neither holds $s_{KM} = 4s_{klmn}$. Finally, for the piezoelectric constants, $d_{kL} = d_{klm}$ if $l = m$, and if $l <> m$ then $d_{kL} = 2d_{klm}$. The matrices corresponding to the strain, compliances, and piezoelectric constants can then be denoted S, s and d, respectively. The factors of 2 in the above definitions are introduced in order that the matrix forms of the linear constitutive relations should be free from such factors. The tensor forms of the constitutive equations are

$$t_{kl} = c_{klmn}S_{mn} - e_{jkl}E_j$$
$$D_k = e_{klm}S_{lm} + \epsilon_{kl}E_l$$

or alternatively

$$S_{kl} = s_{klmn}t_{mn} + d_{jkl}E_j$$
$$D_k = d_{klm}t_{lm} + \epsilon_{kl}E_l$$

and in matrix form with the above definitions the corresponding equations are

$$T = cS - \tilde{e}E$$
$$D = eS + \epsilon E \tag{A5.17}$$

and

$$S = sT + \tilde{d}E$$
$$D = dT + \epsilon E \tag{A5.18}$$

where it should be noted that ϵ has been used to denote the matrix of dielectric constants in both 'clamped' and 'free' cases. The numerical values of the matrix elements are of course different in the two cases.

Having established the reduced notation, it remains to determine how the matrix representations of the third- and fourth-order tensor properties transform under changes of the coordinate system. Under the coordinate transformation Eqn (A5.1), the tensor stresses transform according to

$$t'_{km} = a_{jk}a_{lm}t_{jl} \tag{A5.19}$$

By writing this and the corresponding expression for the transformation of the tensor strains out in full, and replacing the tensor components by matrix components throughout, the transformations for the matrix stresses and strains can be obtained by inspection in the form

$$T' = \tilde{M}T \qquad S' = \tilde{N}S \tag{A5.20}$$

where M and N are 6×6 matrices whose elements are formed from the elements of the transformation matrix A. Their explicit representations are given in Figs. 5.3 and 5.4. The inverse transformations are obtained in the same way from the inverse of A, but since A^{-1} is just the transpose of A, the inverses of M and N can be obtained by simply reversing all pairs of indices for the elements of A in Figs. A5.3 and A5.4. It then follows by inspection that

$$\tilde{N}^{-1} = M \tag{A5.21}$$

so that Eqns (A5.20) can be replaced by

$$T' = \tilde{M}T \qquad S = MS' \tag{A5.22}$$

With the transformation laws for the electric field and displacement E and D in the matrix form

$$E = AE' \qquad D' = \tilde{A}D$$

the constitutive relations Eqn (A5.17) can then be manipulated into the form

$$T' = (\tilde{M}cM)S' - (\tilde{M}\tilde{e}A)E'$$
$$D' = (\tilde{A}eM)S' + (\tilde{A}\epsilon A)E'$$

a_{11}^2	a_{12}^2	a_{13}^2	$a_{12}a_{13}$	$a_{13}a_{11}$	$a_{11}a_{12}$
a_{21}^2	a_{22}^2	a_{23}^2	$a_{22}a_{23}$	$a_{23}a_{21}$	$a_{21}a_{22}$
a_{31}^2	a_{32}^2	a_{33}^2	$a_{32}a_{33}$	$a_{33}a_{31}$	$a_{31}a_{32}$
$2a_{21}a_{31}$	$2a_{22}a_{32}$	$2a_{23}a_{33}$	$a_{22}a_{33}+a_{32}a_{23}$	$a_{23}a_{31}+a_{33}a_{21}$	$a_{21}a_{32}+a_{31}a_{22}$
$2a_{31}a_{11}$	$2a_{32}a_{12}$	$2a_{33}a_{13}$	$a_{32}a_{13}+a_{12}a_{33}$	$a_{33}a_{11}+a_{13}a_{31}$	$a_{31}a_{12}+a_{11}a_{32}$
$2a_{11}a_{21}$	$2a_{12}a_{22}$	$2a_{13}a_{23}$	$a_{12}a_{23}+a_{22}a_{13}$	$a_{13}a_{21}+a_{23}a_{11}$	$a_{11}a_{22}+a_{21}a_{12}$

Fig. A5.3 Matrix M.

a_{11}^2	a_{12}^2	a_{13}^2	$2a_{12}a_{13}$	$2a_{13}a_{11}$	$2a_{11}a_{12}$
a_{21}^2	a_{22}^2	a_{23}^2	$2a_{22}a_{23}$	$2a_{23}a_{21}$	$2a_{21}a_{22}$
a_{31}^2	a_{32}^2	a_{33}^2	$2a_{32}a_{33}$	$2a_{33}a_{31}$	$2a_{31}a_{32}$
$a_{21}a_{31}$	$a_{22}a_{32}$	$a_{23}a_{33}$	$a_{22}a_{33}+a_{32}a_{23}$	$a_{23}a_{31}+a_{33}a_{21}$	$a_{21}a_{32}+a_{31}a_{22}$
$a_{31}a_{11}$	$a_{32}a_{12}$	$a_{33}a_{13}$	$a_{32}a_{13}+a_{12}a_{33}$	$a_{33}a_{11}+a_{13}a_{31}$	$a_{31}a_{12}+a_{11}a_{32}$
$a_{11}a_{21}$	$a_{12}a_{22}$	$a_{13}a_{23}$	$a_{12}a_{23}+a_{22}a_{13}$	$a_{13}a_{21}+a_{23}a_{11}$	$a_{11}a_{22}+a_{21}a_{12}$

Fig. A5.4 Matrix N.

It then follows that the desired transformation laws are

$$c' = \widetilde{M}cM \qquad e' = \widetilde{A}eM \qquad (A5.23)$$

Similar expressions can be derived for the compliances s and the d piezoelectric constants if desired.

When the transformation matrix A corresponds to a symmetry operation of the material, then the Eqns (A5.23) become identities that have to be satisfied by the material constants

$$c = \tilde{M}cM \qquad e = \tilde{A}eM \qquad (A5.24)$$

These are exactly analogous to the identities (Eqn (A5.14)) already discussed in detail for the second-rank tensor case, differing only in being more tedious to apply. Consequently, only the results of applying the symmetry operations of quartz to the arrays of elastic and piezoelectric constants are given here. It is found that the number of independent constants reduces to two piezoelectric and six elastic constants, with the matrices having the following forms.

Piezoelectric constants

$$\begin{pmatrix} e_{11} & -e_{11} & 0 & e_{14} & 0 & 0 \\ 0 & 0 & 0 & 0 & -e_{14} & -e_{11} \\ 0 & 0 & 0 & 0 & 0 & 0 \end{pmatrix}$$

Elastic constants

$$\begin{pmatrix} c_{11} & c_{12} & c_{13} & c_{14} & 0 & 0 \\ c_{12} & c_{11} & c_{13} & -c_{14} & 0 & 0 \\ c_{13} & c_{13} & c_{33} & 0 & 0 & 0 \\ c_{14} & -c_{14} & 0 & c_{44} & 0 & 0 \\ 0 & 0 & 0 & 0 & c_{44} & c_{14} \\ 0 & 0 & 0 & 0 & c_{14} & c_{66} \end{pmatrix}$$

where

$$c_{66} = (c_{11} - c_{12})/2$$

A5.5 TRANSFORMATION EQUATIONS FOR ROTATED Y-CUTS

The plate coordinate system for a rotated Y-cut (Chapter 2) is obtained from the crystallographic axes by rotation about Ox_1 through an angle θ. Hence the A matrix has the form of Eqn (A5.5). The corresponding M matrix is shown in Fig. A5.5. Using these matrices in the transformations of Eqn (A5.12) and

1	0	0	0	0	0
0	c^2	s^2	$-cs$	0	0
0	s^2	c^2	cs	0	0
0	$2cs$	$-2cs$	c^2-s^2	0	0
0	0	0	0	c	s
0	0	0	0	$-s$	c

Fig. A5.5 Matrix M for rotated Y-cuts.

(A5.23), along with the matrices for the material constants obtained by applying the symmetry identities, leads to expressions for the material constants in the plate coordinate system. Once again, the algebra is tedious but straightforward, so only the results are given.

Dielectric constants

$\epsilon'_{11} = \epsilon_{11}$
$\epsilon'_{22} = c^2\epsilon_{11} + s^2\epsilon_{33}$
$\epsilon'_{33} = s^2\epsilon_{11} + c^2\epsilon_{33}$
$\epsilon'_{23} = cs(\epsilon_{33} - \epsilon_{11})$

Piezoelectric constants

$e'_{11} = e_{11}$
$e'_{12} = -c^2 e_{11} + 2cse_{14}$
$e'_{13} = -s^2 e_{11} - 2cse_{14}$
$e'_{14} = cse_{11} + (c^2 - s^2)e_{14}$
$e'_{25} = -c^2 e_{14} + sce_{11}$

$$e'_{26} = -cse_{14} - c^2 e_{11}$$
$$e'_{35} = sce_{14} - s^2 e_{11}$$
$$e'_{36} = s^2 e_{14} + sce_{11}$$

Elastic constants

$$c'_{11} = c_{11}$$
$$c'_{22} = c^4 c_{11} + s^4 c_{33} + 2s^2 c^2 (2c_{44} + c_{13}) - 4sc^3 c_{14}$$
$$c'_{33} = s^4 c_{11} + c^4 c_{33} + 2s^2 c^2 (2c_{44} + c_{13}) + 4cs^3 c_{14}$$
$$c'_{44} = c_{44} + s^2 c^2 (c_{11} + c_{33} - 2c_{13} - 4c_{44}) + 2cs(c^2 - s^2) c_{14}$$
$$c'_{55} = c^2 c_{44} - 2scc_{14} + s^2 c_{66}$$
$$c'_{66} = s^2 s_{44} + 2scc_{14} + c^2 c_{66}$$
$$c'_{12} = c^2 c_{12} + s^2 c_{13} + 2csc_{14}$$
$$c'_{13} = s^2 c_{12} + c^2 c_{13} - 2csc_{14}$$
$$c'_{14} = (c^2 - s^2) c_{14} - cs(c_{12} - c_{13})$$
$$c'_{23} = (c^4 + s^4) c_{13} + c^2 s^2 (c_{11} + c_{33} - 4c_{44}) + 2cs(c^2 - s^2) c_{14}$$
$$c'_{24} = c^2 (4s^2 - 1) c_{14} + sc[s^2 c_{33} - c^2 c_{11} + (2c_{44} + c_{13})(c^2 - s^2)]$$
$$c'_{34} = -s^2 (4c^2 - 1) c_{14} + sc[c^2 c_{33} - s^2 c_{11} - (2c_{44} + c_{13})(c^2 - s^2)]$$
$$c'_{56} = (c^2 - s^2) c_{14} - sc(c_{66} - c_{44})$$

In all the above, $c = \cos(\theta)$ and $s = \sin(\theta)$, and all constants not explicitly listed are either determined by the symmetry of the arrays or else are zero.

Appendix 6
Wave propagation in isotropic plates

A6.1 FIELD EQUATIONS AND CONSTITUTIVE RELATIONS

In an isotropic material there is of necessity no piezoelectricity. Hence the field equations and constitutive relations of the linear theory set out in Appendix 4 reduce (in the absence of body forces) to

$$t_{kl,l} = \rho \ddot{u}_k$$
$$t_{kl} = c_{klmn} S_{mn} \tag{A6.1}$$
$$S_{mn} = (u_{m,n} + u_{n,m})/2$$

where t_{kl} and S_{kl} are the stress and strain tensors, u_k is the mechanical displacement, ρ is the density and the c_{klmn} are the elastic constants. Since in an isotropic material all directions and coordinate systems are equivalent, every coordinate transformation is equivalent to a symmetry operation in the sense of Appendix 5. Therefore the elastic constants must satisfy the identities

$$c_{ikmp} = a_{ij} a_{kl} a_{mn} a_{pq} c_{jlnq}$$

for all orthogonal transformations a_{kl}, which implies that, using the reduced matrix notation of Appendix 5, the matrix of elastic constants must have the simple form

$$\begin{pmatrix} c_{11} & c_{12} & c_{12} & 0 & 0 & 0 \\ c_{12} & c_{11} & c_{12} & 0 & 0 & 0 \\ c_{12} & c_{12} & c_{11} & 0 & 0 & 0 \\ 0 & 0 & 0 & c_{44} & 0 & 0 \\ 0 & 0 & 0 & 0 & c_{44} & 0 \\ 0 & 0 & 0 & 0 & 0 & c_{44} \end{pmatrix}$$

where the additional relation

$$c_{11} = c_{12} + 2c_{44} \tag{A6.2}$$

must also hold.

A6.2 PLANE WAVE SOLUTIONS

As all directions are equivalent in an isotropic material, in considering the propagation of plane waves there is no loss of generality in assuming that the propagation direction is along the x_2 axis. Then in the notation of Section 2.1, the wave normal n_k has components (0,1,0). Taking account of the effect of isotropy on the elastic constants, it then follows from Eqn (2.11) that the tensor L_{km} reduces to diagonal form with $L_{11} = c_{44} = L_{33}$ and $L_{22} = c_{11}$. Consequently Eqn (2.9) separates into three uncoupled wave equations each involving a single displacement component

$$c_{44}u_{1,22} = \rho \ddot{u}_1$$
$$c_{11}u_{2,22} = \rho \ddot{u}_2 \qquad (A6.3)$$
$$c_{44}u_{3,22} = \rho \ddot{u}_3$$

The first and third of these represent transverse or shear waves propagating along x_2 with a phase velocity V_s, whereas the second represents a longitudinal or *dilatational* wave with velocity V_d, where

$$V_s = (c_{44}/\rho)^{1/2} \qquad V_d = (c_{11}/\rho)^{1/2} \qquad (A6.4)$$

Since from Eqn (A6.2) $c_{11} > c_{44}$ for positive c_{12}, the longitudinal wave velocity V_d is greater than the shear wave velocity V_s.

In each of the three waves described by Eqns (A6.3), there is only a single non-zero strain, S_6, S_2 and S_4, respectively. In the transverse waves, the corresponding non-zero stresses are T_6 and T_4, respectively, while in the longitudinal wave, all three extensional stress components T_1, T_2 and T_3 are non-zero but there are no shear stresses.

A6.3 BOUNDARY CONDITIONS FOR A PLATE

Consider a plate in the XZ plane with major surfaces at $y = x_2 = \pm h$, so that the plate thickness is $2h$. If the major surfaces are stress free, then the appropriate boundary conditions are that $t_{2k} = 0$ when $y = \pm h$, or in the reduced notation, that $T_6 = T_2 = T_4 = 0$ on the major surfaces. Because of the form of the elastic constant matrix, the vanishing of the shear stresses T_6 and T_4 implies the vanishing of the corresponding shear strains S_6 and S_4, while the vanishing of T_2 implies

$$c_{12}S_1 + c_{11}S_2 + c_{12}S_3 = 0$$

In terms of the displacements $u = u_1$, $v = u_2$ and $w = u_3$, the boundary conditions may be written

$$v_{,z} + w_{,y} = 0$$

$$u_{,y} + v_{,x} = 0 \tag{A6.5}$$
$$c_{11}v_{,y} + c_{12}(u_{,x} + w_{,z}) = 0$$

where x, y, z has been written for x_1, x_2, x_3 and Eqns (A6.5) hold at both major surfaces, $y = \pm h$.

A6.4 THICKNESS MODES

As discussed in Section 2.2, pure thickness modes in plates can be regarded as standing wave systems resulting from the reflection at the major surfaces of plane waves propagating along the thickness. In the isotropic case, the analysis is very simple compared to the anisotropic piezoelectric case discussed in Chapter 2. There are just two families of thickness modes, one resulting from the reflections of transverse waves and termed the *thickness shear* or TS family, the other resulting from reflections of longitudinal waves and termed the *thickness extensional* or TE family.

In the TS case, the appropriate solution of the first of Eqns (A6.3) is, assuming a harmonic time dependence and writing u for u_1, y for x_2

$$u = [A\sin(ky) + B\cos(ky)]\exp(j\omega t) \tag{A6.6}$$

where k is the wave number along the thickness, and A, B are constants to be chosen to satisfy the boundary conditions. In order for Eqn (A6.6) to satisfy the wave equation, k and ω must satisfy

$$k^2 = \omega^2/V_s^2 \tag{A6.7}$$

The appropriate boundary condition is the second of Eqns (A6.5), which in this case reduces to $u_{,y} = 0$ at $y = \pm h$. Hence

$$kA\cos(kh) - kB\sin(kh) = 0$$
$$kA\cos(kh) + kB\sin(kh) = 0$$

Adding and subtracting leads to two sets of conditions for symmetric modes and anti-symmetric modes:

Symmetric case: $A = 0$, $\quad \sin(kh) = 0$ \hfill (A6.8)

Anti-symmetric: $B = 0$, $\quad \cos(kh) = 0$ \hfill (A6.9)

The conditions on (kh) are equivalent to the single requirement that $\sin(2kh) = 0$, in turn equivalent to

$$k = N\pi/2h \tag{A6.10}$$

with N integral. This then leads through Eqn (A6.7) to the frequency equation

$$\omega = (N\pi/2h)V_s \tag{A6.11}$$

which is essentially Eqn (2.42). In the frequency equation, odd N values correspond to the anti-symmetric modes, even N values to the symmetric modes.

Precisely the same analysis holds for the TE case provided that the displacement v and phase velocity V_d are used in place of u and V_s. The frequency equation for TE modes is then

$$\omega = (N\pi/2h)V_d \qquad (A6.12)$$

A6.5 THICKNESS TWIST (TT) AND FACE SHEAR (FS) WAVES

Consider a transverse wave with particle displacement u along x, propagating in a direction in the YZ plane. If the wave number is k and the unit vector along the direction of propagation has the components $(0, n_y, n_z)$, then for an angular frequency ω an appropriate form for u is

$$u = A\exp\{j[\omega t - k(n_y y + n_z z)]\}$$

where $k^2 = \omega^2/V_s^2$ as before.

A similar wave propagating in the direction $(0, -n_y, n_z)$ would then have the corresponding form

$$u = B\exp\{j[\omega t - k(-n_y y + n_z z)]\}$$

Superimposing these two waves then gives rise to a displacement u

$$u = [A\exp(-jkn_y y) + B\exp(+jkn_y y)]\exp(j\omega t - jkn_z z)$$

or equivalently

$$u = [A'\sin(kn_y y) + B'\cos(kn_y y)]\exp(j\omega t - jkn_z z) \qquad (A6.13)$$

This represents a travelling wave in the z direction, with wave number $k_z = kn_z$, a corresponding phase velocity $V = \omega/k_z$, and an amplitude varying with y. Since u depends on both y and z, the non-zero strains are the shear strains S_5 and S_6 and the non-zero stresses the shear stresses T_5 and T_6. From Section A6.3, if $u_{,y}$ can be made to vanish at $y = \pm h$, then Eqn (A6.13) represents a plate wave solution. But the y dependence of u in Eqn (6.13) is just that of u in Eqn (A6.6) provided k is replaced by kn_y, so that the boundary conditions will be satisfied if

$$k_y = kn_y = N\pi/2h \qquad (A6.14)$$

The *dispersion relation* Eqn (A6.7) then becomes

$$k^2 = k_y^2 + k_z^2 = \omega^2/V_s^2$$

or

$$w^2/V_s^2 = (N\pi/2h)^2 + k_z^2 \qquad (A6.15)$$

where as in the thickness mode case, even N corresponds to modes symmetric in y, odd N to modes anti-symmetric in y.

This can be put into a normalized form by multiplying by $(2h/\pi)^2$ and defining

$$\zeta = 2hk_z/\pi$$
$$\Omega = \omega/\omega_1 \qquad\qquad\qquad (A6.16)$$
$$\omega_1 = (\pi/2h)V_s$$

The normalized dispersion relation is then

$$\Omega^2 = \zeta^2 + N^2 \qquad\qquad\qquad (A6.17)$$

The character of the waves described by Eqn (A6.13) is quite different for the case $N = 0$ and $N > 0$. In the latter case, the particle motion in the wave is, as illustrated in Fig. A6.1, a twisting motion along the direction of propagation: hence these waves are called *thickness twist* or TT waves. In the case $N = 0$, the displacement u is independent of y, and the deformation is a shear in the plane of the plate. Thus the wave is termed a *face shear* or FS wave. Apart from these differences in the particle displacements, the crucial difference between the TT and FS waves lies in the existence of a cut-off frequency for each TT wave.

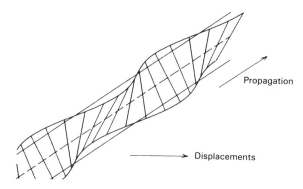

Fig. A6.1 Particle displacements for TT waves.

Analytically, it follows from the normalized dispersion relation that for $N > 0$, the normalized wavenumber ζ is imaginary for all frequencies such that $\Omega < N$. This implies an imaginary k_z, which on substitution into Eqn (A6.13) shows that u decays exponentially with z. Thus for normalized frequencies $\Omega < N$, Eqn (A6.13) represents an *evanescent* rather than a travelling wave. The cut-off frequencies are just the thickness mode frequencies, and in terms of the physical picture of the plate wave as being built up by successive reflections of plane waves at the surfaces of the plate, cut-off corresponds to the limiting case when the plane waves are normally incident on the plate surfaces. As discussed in Chapter 3, as this condition is approached, the phase velocity of the plate wave tends to infinity, while its group velocity tends to zero. The phenomenon of cut-off in elastic wave propagation is closely analogous to the cut-off phenomenon in electromagnetic waveguides.

In the special case $N = 0$, the displacement u depends only on z, so the only strain generated is S_5 and the only stress T_5. But T_5 does not contribute to the surface tractions t_{2k} so that in this case the boundary conditions for the plate are trivially satisfied. Then Eqn (A6.13) actually represents a plane wave solution, and the plate wave velocity coincides with the bulk shear wave velocity.

Figure A6.2 shows a plot of the branches of the dispersion relation Eqn (A6.17) for both real and imaginary $k_z(\zeta)$. In the imaginary case, is set equal to $j\bar{\zeta}$, and $\bar{\zeta}$ is plotted to the left of the frequency axis in place of negative ζ. It should be noted that the slope of the TT branches at $\zeta = 0$ is zero, corresponding to the limiting value of zero for the group velocity of these branches.

A6.6 THE RAYLEIGH FREQUENCY EQUATIONS FOR AN ISOTROPIC PLATE

In the previous section, TT waves were built up by combining two bulk plane waves, each with particle displacements in the plane of the plate. If a transverse wave propagating in the XY plane is considered, say along the direction $(n_x, n_y, 0)$, and if the particle displacement is also in the XY plane, then the displacement will have components u and v along both x and y. If the

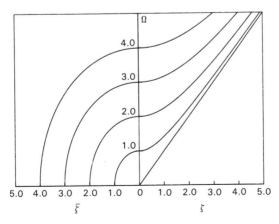

Fig. A6.2 Dispersion relations for TT and FS waves.

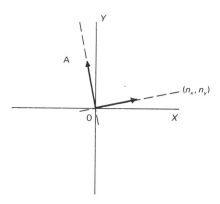

Fig. A6.3 Transverse wave in OXY plane.

amplitude of the wave is A, then from Fig. A6.3 the amplitudes of the u and v components will be respectively $-An_y$ and An_x. If k is the wavenumber, then the general expression for the displacements will be

$$u = -An_y \exp[j\omega t - jk(n_x x + n_y y)]$$
$$v = An_x \exp[j\omega t - jk(n_x x + n_y y)] \tag{A6.18}$$

where, because of the transverse nature of the wave, the dispersion relation $k^2 = \omega^2/V_s^2$ holds.

Just as in the previous case of TT waves, a second transverse wave of amplitude B and travelling in the direction $(n_x, -n_y, 0)$ can be superimposed on the first to give a resultant wave with displacements of the form

$$u = n_y\{-A\exp(-jkn_y y) + B\exp(+jkn_y y)\}\exp(j\omega t - jk_x x)$$
$$v = n_x\{+A\exp(-jkn_y y) + B\exp(+jkn_y y)\}\exp(j\omega t - jk_x x) \tag{A6.19}$$

where $k_x = kn_x$. As before, this is in the form of a travelling wave along the plate, in this case in the x direction, with wavenumber k_x and phase velocity $V = \omega/k_x$, and with an amplitude depending on y.

Two adjustable constants A and B are available to match the boundary conditions. However, whereas in the TT case the boundary conditions reduced to the vanishing of $u_{,y}$ at $y = \pm h$, in this case with both u and v depending on x and y, the conditions to be satisfied are, from Eqns (A6.5)

$$u_{,y} + v_{,x} = 0$$
$$c_{11}v_{,y} + c_{12}u_{,x} = 0 \tag{A6.20}$$

Since these have to be satisfied at $y = \pm h$, there are four conditions to satisfy with only two adjustable parameters. The extra parameters required are obtained by also considering a combination of longitudinal waves. Consider then a longitudinal wave, amplitude C, propagating along the direction

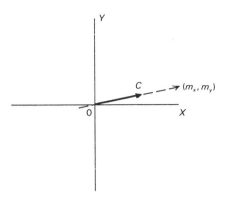

Fig. A6.4 Longitudinal wave in OXY plane.

$(m_x, m_y, 0)$ with wavenumber l. From Fig. A6.4, the u and v displacements will respectively have amplitudes Cm_x and Cm_y, so that

$$u = Cm_x \exp[j\omega t - jl(m_x x + m_y y)]$$
$$v = Cm_y \exp[j\omega t - jl(m_x x + m_y y)] \tag{A6.21}$$

where $l^2 = \omega^2/V_d^2$. Superimposing a second similar wave with amplitude D and direction $(m_x, -m_y, 0)$ then gives in analogy to Eqn (A6.19)

$$u = m_x\{C\exp(-jlm_y y) + D\exp(+jlm_y y)\}\exp(j\omega t - jl_x x)$$
$$v = m_y\{C\exp(-jlm_y y) - D\exp(+jlm_y y)\}\exp(j\omega t - jl_x x) \tag{A6.22}$$

where $l_x = lm_x$.

In order for these expressions to represent a travelling wave along x with the same phase velocity as that represented by Eqns (A6.19), it is necessary to require that the wavenumbers along x match, that is

$$k_x = kn_x = l_x = lm_x \tag{A6.23}$$

Since l and k are known in terms of the frequency and the bulk wave velocities V_d and V_s, this determines $k_y = kn_y$ and $l_y = lm_y$ via

$$k_x^2 + k_y^2 = \omega^2/V_s^2$$
$$k_x^2 + l_y^2 = \omega^2/V_d^2 \tag{A6.24}$$

If now the sum of the expressions in Eqns (A6.19) and (A6.22) is substituted into the boundary conditions Eqns (A6.20), a set of four homogeneous linear equations in the unknowns A, B, C and D result. The necessary condition for non-trivial solutions to exist is that the determinant of the coefficients of this system of equations vanishes; the vanishing of the determinant in turn leads to a transcendental equation for the wave numbers k_y, l_y which is analogous to the simple trigonometric condition $\sin(2k_y h) = 0$ in the TT case.

The set of four simultaneous equations can be reduced to two independent pairs by the device of adding and subtracting the boundary conditions that apply at $y = h$ and at $y = -h$. This eliminates in turn the anti-symmetric and the symmetric parts of the solutions, and by straightforward but tedious algebra leads to the equations

$$(A + B)\cos(k_y h)(k_y n_y - k_x n_x) = \\ (C - D)\cos(l_y h)(l_y m_x + k_x m_y)$$
$$(A + B)\sin(k_y h)(c_{12} k_x n_y - c_{11} k_y n_x) = \\ (C - D)\sin(l_y h)(c_{12} k_x m_x + c_{11} l_y m_y)$$
(A6.25)

and

$$(A - B)\sin(k_y h)(k_y n_y - k_x n_x) = \\ (C + D)\sin(l_y h)(l_y m_x + k_x m_y)$$
$$(A - B)\cos(k_y h)(c_{12} k_x n_y - c_{11} k_y n_x) = \\ (C + D)\cos(l_y h)(c_{12} k_x m_x + c_{11} l_y m_y)$$
(A6.26)

To exploit the symmetry of the solutions, first set $A = B$ and $D = -C$, so that Eqns (A6.26) are satisfied trivially, and Eqns (A6.25) reduce to a pair of equations for A and C. The solutions for u and v then become

$$u = \{2jn_y A \sin(k_y y) - 2jm_x C\sin(l_y y)\}\exp(j\omega t - jk_x x)$$
$$v = \{2n_x A\cos(k_y y) + 2m_y C\cos(l_y y)\}\exp(j\omega t - jk_x x)$$
(A6.27)

with amplitude ratios determined from Eqns (A6.25), and the transcendental equation that follows from setting the determinant of coefficients in Eqns (A6.25) to zero

$$\tan(l_y h)/\tan(k_y h) = -4k_y l_y k_x^2/(k_y^2 - k_x^2)^2$$
(A6.28)

This family of solutions is termed the *anti-symmetric* family of motions of the isotropic plate. The symmetric family is obtained by setting $B = -A$ and $C = D$ and using Eqns (A6.26) to determine the ratio C/A. The condition for non-trivial solutions differs from Eqn (A6.28) only in the reversal of the two tangent functions,

$$\tan(k_y h)/\tan(l_y h) = -4k_y l_y k_x^2/(k_y^2 - k_x^2)^2$$
(A6.29)

and the displacements take the form

$$u = \{-2n_y A \cos(k_y y) + 2m_x C\cos(l_y y)\}\exp(j\omega t - jk_x x)$$
$$v = \{-2jn_x A\sin(k_y y) - 2jm_y C\sin(l_y y)\}\exp(j\omega t - jk_x x)$$
(A6.30)

The results expressed by Eqns (A6.28) and (A6.29) were first obtained by Rayleigh (1889) and are known as the Rayleigh frequency equations. The preceding analysis provides the basic framework for a discussion of the various types of wave that can propagate in an isotropic plate, subject only to the restriction that the particle displacement be in the plane defined by the direction of propagation and the plate normal. Unfortunately, the complex form of the frequency equations and the resulting large variety of solutions make anything approaching a complete discussion out of the question. A more detailed treatment, along with references to the large literature on this subject, may be found in Tiersten (1969). For present purposes, a brief discussion of the anti-symmetric family of solutions that contain as limiting cases the shear thickness modes will suffice.

A6.7 ANTI-SYMMETRIC SOLUTIONS: THICKNESS SHEAR (TS) AND FLEXURAL (F) WAVES

As in the discussion of TT modes, it is convenient to work in terms of normalized quantities, with wavenumbers expressed in terms of the number of half wavelengths that would be contained in a distance equal to the thickness of the plate, and frequencies in terms of the fundamental thickness shear frequency. Define

$$
\begin{aligned}
\omega_1 &= \pi V_s/2h \\
\Omega &= \omega/\omega_1 \\
\xi &= 2hk_x/\pi \\
\eta_{(1)} &= 2hk_y/\pi \\
\eta_{(2)} &= 2hl_y/\pi
\end{aligned}
\tag{A6.31}
$$

Then Eqns (A6.24) can be written

$$
\begin{aligned}
\xi^2 + \eta_{(1)}^2 &= \Omega^2 \\
\xi^2 + \eta_{(2)}^2 &= \Omega^2 (V_s/V_d)^2
\end{aligned}
\tag{A6.32}
$$

and the frequency equation (A6.28) becomes

$$
\tan(\pi\eta_{(2)}/2)/\tan(\pi\eta_{(1)}/2) = -4\eta_{(1)}\eta_{(2)}\xi^2/(\eta_{(1)}^2 - \xi^2)^2
\tag{A6.33}
$$

The general problem is then to determine the relationship between the normalized frequency Ω and the normalized wavenumber ξ in a manner analogous to the determination of the simple Ω versus ζ relationship for TT modes shown in Fig. A6.2. In principle this can be done by substituting for $\eta_{(1)}$ and $\eta_{(2)}$ in Eqn (A6.33) using eqns (A6.32) and solving the resultant equation for Ω in terms of ξ. However, in practice this is far from straightforward.

For present purposes, the complete spectrum is not in any case required. On physical grounds, since the thickness modes already give a reasonable insight into resonator behaviour, it is to be expected that taking into account only slight departures from the pure thickness modes will add considerably to the accuracy of the analysis. Of major interest then is the behaviour of those

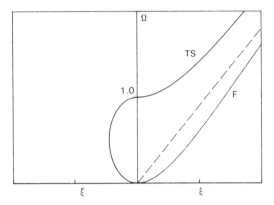

Fig. A6.5 Dispersion relations for TS and F waves.

branches of the spectrum that tend to pure thickness shear modes as the wavenumber ξ tends to zero. In addition, some knowledge of the lowest, flexural mode, branch of the dispersion relation provides an insight into the unwanted coupling of the thickness shear modes in resonators with flexure modes.

Sufficient information for a qualitative understanding of this latter point can be obtained by considering the behaviour of the frequency equation in the limit of both small wavenumbers and low frequencies. This yields the classical dispersion relation for flexural waves in thin plates. This can then be matched to the asymptotic behaviour when both frequency and wavenumber are large, it turning out that the asymptotic phase velocity in this limit is just that of Rayleigh surface waves (Tiersten, 1969). This process is discussed in detail by Tiersten, and the shape of the resulting flexural mode branch is shown in Fig. A6.5.

In the case of the thickness shear modes, the approximate analysis begins by noting that the appropriate thickness mode frequencies do follow from Eqns (A6.32) and (A6.33) in the limit $\xi = 0$. For $\xi = 0$, either $\sin(\pi\eta_{(2)}/2) = 0$ or $\cos(\pi\eta_{(1)}/2) = 0$. The former choice corresponds to TS modes, the latter to TS modes, and adopting this option leads to the conditions

$$\eta_{(1)} = \Omega = N, \quad N = 1, 3, 5, \ldots.$$

just as in the previous analysis of thickness modes. Considering now modes that only differ marginally from pure thickness modes, set

$$\begin{aligned}\Omega &= N(1 + \epsilon) \\ \eta_{(1)} &= N(1 + \delta)\end{aligned} \quad (A6.34)$$

where both ϵ and δ are $\ll 1$. Then from the first of Eqns (A6.32), to first order in the small quantities

$$\delta = \epsilon - \xi^2/2N^2 \quad (A6.35)$$

From the second of Eqns (A6.32), again to a first approximation

$$\eta_{(2)} = \kappa N \tag{A6.36}$$

where κ has been written for the ratio of shear and dilatational bulk wave velocities

$$\kappa = V_s/V_d \tag{A6.37}$$

With the aid of these approximations, and making use of the expansion

$$\cot(\pi N(1 + \delta)/2) = -\tan(\pi N\delta/2) \sim -\pi N\delta/2 \tag{A6.38}$$

the frequency equation (A6.33) takes on the approximate form

$$\epsilon = \{1 + 16\kappa \cot(\pi \kappa N/2)/\pi N\} \xi^2/2N^2 \tag{A6.39}$$

This allows the shape of the thickness shear branches near the axis to be sketched for both real and imaginary values of ξ. The appropriate portions of the dispersion curves are shown in Fig. A6.5, along with the lowest flexural mode branch already discussed. It should be noted that as required from physical considerations, the slope of the curves vanishes at $\xi = 0$, corresponding to zero group velocity at the cut-off frequencies.

References

Ballato, A. (1970). Resonance in Piezoelectric Vibrators. *Proc. IEEE*, **58**, 149-51.
Ballato, A. (1977). Doubly Rotated Thickness Mode Plate Vibrators. In *Physical Acoustics: Principles and Methods*, Vol XIII (W.P. Mason and R.N. Thurston, eds). Academic Press.
Bechmann, R. (1955). Influence of the Order of Overtone on the Temperature Coefficient of Frequency of AT-Type Quartz Resonators. *Proc. IRE*, **43**, 1667-8.
Bechamnn, R. (1956). Frequency-Temperature-Angle Characteristics of AT-Type Resonators Made of Natural and Synthetic Quartz. *Proc. IRE*, **44**, 1600-7.
Bechmann, R. (1960). Frequency-Temperature-Angle Characteristics of AT- and BT-Type Quartz Oscillators in an Extended Temperature Range. *Proc. IRE*, **48**, 1494.
Bechmann, R. (1961). Thickness shear mode quartz cut with small second and third order temperature coefficients (RT cut). *Proc. IRE*, **49**, 1454.
Bernstein, M. (1967). Increased Crystal Unit Resistance at Oscillator Noise Levels. *Proc. 21st Ann. Freq. Control Symposium*, US Army Electronics Command, Ft. Monmouth, N.J., 244-58.
Besson, R. (1976). A New Piezoelectric Resonator Design. *Proc. 30th Ann. Freq. Control Symposium*, US Army Electronics Command, Ft. Monmouth, N.J., 78-83.
Bond, W.L. (1976). *Crystal Technology*. Wiley, New York.
Bottom, V.E. (1982). *Introduction to Quartz Crystal Unit Design*. Van Nostrand Reinhold, New York.
Bottom, V. and Ives, W.R. (1951). U.S. Patent No. 2,743,144.
Brice, J.C. (1985). Crystals for quartz resonators. *Reviews of Modern Physics*, **57**, 105-146.
Buchanan, J.P. (1956). *Handbook of Piezoelectric Crystals for Radio Equipment Designers*. WADC Technical Report 56-156. National Technical Information Service, U.S. Department of Commerce, Washington, D.C.
Cady, W.G. (1964). *Piezoelectricity*. Dover, New York.
EerNisse, E.P. (1975). Quartz Resonator Frequency Shifts Arising from Electrode Stress. *Proc. 29th Ann. Freq. Control Symposium*, US Army Electronics Command, Ft. Monmouth, N.J., 1-4.
Frerking, M.E. (1978). *Crystal Oscillator Design and Temperature Compensation*. Van Nostrand Reinhold, New York.
Gagnepain, J.J. (1981). Nonlinear Properties of Quartz Crystal and Crystal Resonators. *Proc. 35th Ann. Freq. Control Symposium*, US Army Electronics Command, Ft. Monmouth, N.J., 14-30.
Gagnepain, J.J. and Besson, R. (1975). Nonlinear Effects in Piezoelectric Quartz Crystals. In *Physical Acoustics: Principles and Methods*, Vol XI (W.P. Mason and R.N. Thurston, eds). Academic Press, New York.
Gerber, E.A. and Ballato, A. (1985). *Precision Frequency Control*. Academic Press, Orlando.
Guttwein, G.K., Lukaszek, T.J. and Ballato, A.D. (1967). Practical Consequences of

Modal Parameter Control in Crystal Resonators. *Proc. 21st Ann. Freq. Control Symposium*, US Army Electronics Command, Ft. Monmouth, N.J., 115-37.

Heising, R.A. (1946). *Quartz Crystals for Electrical Circuits*. Van Nostrand, New York. Reprinted 1978 by Electronic Industries Association, Washington, D.C.

Hight, S.C. and Willard, G.W. (1937). A simplified circuit for frequency substandards, employing a new type of low-frequency, zero-temperature-coefficient crystal. *Proc. IRE*, **25**, 549-63.

Holland, R. (1974a). Nonuniformly Heated Anisotropic Plates. I. Mechanical Distortion and Relaxation [Quartz Resonator]. *IEEE Trans. Sonics & Ulrasonics*, **SU-21**, 171-8.

Holland, R. (1974b). Nonuniformly Heated Anisotropic Plates. II. Frequency Transients in AT and BT Quartz Plates. *Proc. 1974 Ultrasonics Symposium*, 592-8.

Holland, R. and EerNisse, E.P. (1969). *Design of Resonant Piezoelectric Devices*. Research Monograph No. 56, MIT Press, Cambridge, Massachusetts.

Humpherys, D.S. (1970). *The Analysis, Design, and Synthesis of Electrical Filters*. Prentice-Hall, Englewood Cliffs, N.J.

IEC (1973). *Basic method for the measurement of resonance frequency and equivalent series resistance of quartz crystal units by zero phase technique in a pi-network*. IEC Publication 444.

IEC (1980). *Measurement of quartz crystal unit parameters by zero phase technique in a pi-network. Part 2: Phase offset method for measurement of motional capacitance of quartz crystal units*. IEC Publication 444-2.

IEC (1981). *Quartz crystal controlled oscillators. Part 2: Guide to the use of quartz crystal controlled oscillators*. IEC Publication 679-2.

IEEE (1978). *IEEE Standard on Piezoelectricity*. IEEE Std 176-1978.

Kartaschoff, P. (1978). *Frequency and Time*. Academic Press, London.

Kawashima, H., Sato, H. and Ochiai, O. (1980). New Frequency Temperature Characteristics of Miniaturized GT-cut Quartz Resonators. *Proc. 34th Ann. Freq. Control Symposium*, US Army Electronics Command, Ft. Monmouth, N.J., 131-9.

Lack, F.R., Willard, G.W. and Fair, I.E. (1934). Some Improvements in Quartz Crystal Circuit Elements. *Bell System Technical Journal*, **13**, 453-463.

Lagasse, G., Ho, J. and Bloch, M. (1972). Research and Development of a New Type of Crystal — The FC Cut. *Proc. 26th Ann. Freq. Control Symposium*, US Army Electronics Command, Ft. Monmouth, N.J., 148-151.

Mason, W.P. (1940). A New Crystal Plate Designated the 'GT'. *Proc. IRE*, **28**, 220-3.

Mason, W.P. (1950). *Piezoelectric Crystals and Their Application to Ultrasonics*. Van Nostrand Reinhold, New York.

Matthys, R.J. (1983). *Crystal Oscillator Circuits*. Wiley-Interscience, New York.

Meeker, T.R. (1985). Theory and Properties of Piezoelectric Resonators and Waves. In *Precision Frequency Control* (Gerber, E.A. and Ballato, A., eds). Academic Press, Orlando.

Mindlin, R.D. and Spencer, W.J. (1967). Anharmonic, Thickness-Twist Overtones of Thickness-Shear and Flexural Vibrations of Rectangular, AT-Cut Quartz Plates. *J.A.S.A.*, **42**, 1268-77.

Moore, S.C. (1983). Photolithographic Manufacture of Quartz Devices. *Proc. 5th Annual Quartz Crystal Conference*, Kansas. Electronic Industries Association, Washington, D.C., 115, 130.

Neubig, B. (1979). Technical Information: Design of Crystal Oscillator Circuits. Special Issue of *VHF Communications*, **3, 4**, 1979.

Nonaka, S., Yuuki, T. and Hara, K. (1971). The Current Dependency of Crystal Unit

Resistance at Low Drive Level. *Proc. 25th Ann. Freq. Control Symposium*, US Army Electronics Command, Ft. Monmouth, N.J., 139-47.

Ochiai, O., Mashimo, Y. and Tamura, F. (1986). Miniaturized High-Accuracy Crystal Oscillator with Electrically Adjustable Frequency. *Journal of Electronic Engineering*, **24**, 32-7.

Oita, T. (1986). Crystal Clock Oscillators Reduce Power Consumption and Size of Digital Equipment. *Journal of Electronic Engineering*, **24**, 28-31.

Parzen, B. (1983). *Design of Crystal and other Harmonic Oscillators*. Wiley-Interscience, New York.

Peters, R.D. (1976). Ceramic Flatpack Enclosures for Precision Quartz Crystal Units. *Proc. 30th Ann. Freq. Control Symposium*, US Army Electronics Command, Ft. Monmouth, N.J., 224-31.

Phillips, F.C. (1960). *An Introduction to Crystallography*. Longmans, London.

Rayleigh, Lord (1889). On the Free Vibrations of an Infinite Plate of Homogeneous Isotropic Matter. *Proc. London Math. Soc.*, **20**, 225.

Sheahan, D.F. and Johnson, R.A. (1977). *Modern Crystal and Mechanical Filters*, IEEE, New York.

Smythe, R.C. (1974). Intermodulation in Thickness-Shear Resonators. *Proc. 28th Ann. Freq. Control Symposium*, US Army Electronics Command, Ft. Monmouth, N.J., 5-7.

Temes, G.C. and Mitra, S.K. (1973). *Modern Filter Theory and Design*. Wiley-Interscience, New York.

Tiersten, H.F. (1969). *Linear Piezoelectric Plate Vibrations*. Plenum Press, New York.

Tiersten, H.F. (1974). Analysis of Intermodulation in Rotated Y-cut Quartz Thickness-Shear Resonators. *Proc. 28th Ann. Freq. Control Symposium*, US Army Electronics Command, Ft. Monmouth, N.J., 1-4.

Tiersten, H.F. (1975). Analysis of Nonlinear Resonance in Rotated Y-cut Quartz Thickness-Shear Resonators. *Proc. 29th Ann. Freq. Control Symposium*, US Army Electronics Command, Ft. Monmouth, N.J., 49-53.

Tiersten, H.F. and Smythe, R.C. (1979). An analysis of contoured crystal resonators operating in overtones of coupled thickness shear and thickness twist. *J.A.S.A.*, **65**, 1455-60.

Tyler, L.A. (1960). In *Quartz Resonator Handbook: Manufacturing Guide for AT Type Units* (ed. Bennett, R.E.). Union Thermoelectric Division, Niles, Illinois.

Vig, J.R., Cook, C.F., Schwidtal, K., LeBus, J.W. and Hafner, E. (1974). Surface Studies for Quartz resonators. *Proc. 28th Ann. Freq. Control Symposium*, US Army Electronics Command, Ft. Monmouth, N.J., 96-108.

Vig, J.R. and LeBus, J.W. (1975). Further Results on UV cleaning and Ni Electrobonding. *Proc. 29th Ann. Freq. Control Symposium*, US Army Electronics Command, Ft. Monmouth, N.J., 220-29.

Vigoreux, P. and Booth, C.F. (1950). *Quartz Vibrators and their Applications*. H.M.S.O., London.

Ward, R. (1983). Quartz Resonator Mounting and Packaging Requirements and Techniques. *Proc. 5th Annual Quartz Crystal Conference*, Kansas. Electronic Industries Association, Washington, D.C., 162-175.

Weinberg, L. (1962). *Network Analysis and Synthesis*. McGraw-Hill, New York.

Weyl, H. (1939). *The Classical Groups, their Invariants and Representations*. Princeton University Press.

Zverev, A. (1967). *Handbook of Filter Synthesis*. John Wiley, New York.

FURTHER READING FOR APPENDIX 3

Dixon, R.C. and Eringen, A.C. (1965). A Dynamical Theory of Polar Elastic Dielectrics. *Int. J. Engng. Sci.*, **3**, 359.

Eringen, A.C. (1962). *Nonlinear Theory of Continuous Media*. McGraw-Hill, New York.

Eringen, A.C. (1963). On the Foundations of Electroelastostatics. *Int. J. Engng. Sci*, **1**, 127.

Grindlay, J. (1966). The Elastic Dielectric. *Phys. Rev.*, **149**, 637.

Tiersten, H.F. (1971). On the Nonlinear Equations of Thermoelectroelasticity. *Int. J. Engng. Sci.*, **9**, 587.

Toupin, R.A. (1956). The Elastic Dielectric. *J. Rational Mech. Anal.*, **5**, 849.

Toupin, R.A. (1960). Stress Tensors in Elastic Dielectrics. *Arch. Rational Mech. Anal.*, **5**, 440.

Toupin, R.A. (1963). A Dynamical Theory of Elastic Dielectrics. *Int. J. Engng. Sci.*, **1**, 101.

Index

a mode 37
AC-cut 19
Acoustic coupling 153
Acoustic attenuation 32
Acoustic loss 112-3
Activity dips 55, 117
Adjusting to frequency 81-3
Admittance
 circle 100
 of crystal 92-8
 of motional arm 92-6
Ageing 117
 causes 77, 117
 in crystals 117
 in oscillators 131-4
 processes 77, 117
 rates 118
Air-gap electrodes 78
Alpha quartz 3-5
Amplifier 127
Amplitude
 of oscillation 128
 of vibration 48, 153
Analysis
 of crystal equivalent circuit 91-103
 of filter networks 144
 of oscillator circuits 128-131
Angle
 of crystal cut 17-21, 37
 tolerances in manufacture 20-2, 64, 114
Anisotropic material 6
Anti-resonance 28, 92, 95
Anti-symmetric modes 26
Anti-symmetric tensors 164
Approximate plate theories 49
Approximation, quasi-static 180
Array, of scattering centres 64
AT-cut 19, 37-8, 41
Atmosphere, effect on resonator Q 32, 113
Atomic planes 64
Attenuation
 maximum in filter passband 141
 minimum in filter stopband 141
Autoclave 3
Axes
 coordinate 5, 162
 crystallographic 5
 electric 4

 optic 4
 polar 4
 symmetry 4

b mode 37
Bandbreaks 55, 117
Bandpass filters 141
Bandstop filters 141
Bandwidth
 passband 141
 stopband 141
 transition region 141
Bar resonator 11, 14
Base-plating 78-9
BC-cut 19
Bechmann's number 59
Beta quartz 3, 5
Bevelling 58-9
Bimorph resonator 14
Blank diameter 107-9
Blank geometry 59-60, 107
Bonding 81
Boundary conditions
 plate waves 47, 58, 209
 thickness modes 25, 58, 209
Bragg angle 65
Bragg reflection 65
Bravais-Miller indices 65
BT-cut 19, 37, 42
Butterworth filter 155

c mode 37
Capacitance
 coupling 153-4
 motional, C_1 30, 91, 110
 ratio r 30, 92, 96, 111
 shunt 28-30, 91, 110
 static 28-30, 91, 110
Cascaded lattice filter 148
Centre of symmetry 4
Ceramic filters 143
Characteristic frequencies 92-5, 101-2
Chebyshev filters 155
Chemical etching 77
Chemical polishing 77
CI meter 120
Clapp oscillator 130
Cleaning quartz crystals 77, 118

INDEX

Clock oscillator 137
Colpitts oscillator 130
Compensation, of crystal oscillator 137
Compliance, elastic 195, 202
Constants, material 5, 194
  density 3, 5-7
  dielectric 5-7, 195
  elastic 5-7, 194
  piezoelectric 5-7, 194
  temperature coefficients of 6-8, 32
Constitutive relations
  linear 8, 23, 193
  non-linear 188
Contouring 58-9, 74
  bowl 75
  drum 74
Convention, sign 5, 37
Coordinates 162
  rotation 197
  transformation 162
Coupled modes 9, 12, 14-5, 19, 51
Coupled resonators 153
Coupling, acoustic 153
Coupling coefficient, electromechanical 27, 36-7, 40
Crystal resonators, quartz
  AC-cut 19
  AT-cut 19, 37-8, 41
  BC-cut 19
  BT-cut 19, 37, 42
  CT-cut 19
  DT-cut 19
  FC-cut 37-8
  GT-cut 15, 18
  IT-cut 37
  LC-cut 37
  MT-cut 22
  NT-cut 22
  RT-cut 37
  SC-cut 21, 37-8
  X-cut family 9, 11-12, 22
  Y-cut 12-3, 18
Crystal holders 80-1, 85
Crystal symmetries 4, 6, 197
CT-cut 19
Cultured quartz 3, 5
Cut-off frequency 48, 55, 212

Deformation 172
Delay, group 141
Density 3, 5, 7
Dielectric constants 5-7, 195
Diffraction, X-ray 64
Digital TCXO 138
Dipole moment 160-1
Diode, varactor 138, 140
Direct plating 84
Dispersion relations 46-8, 211, 218
Displacement
  electric 23, 180
  particle 23, 190
Double conversion systems 143
Doubly rotated crystal cuts 21, 35
DT-cut 19
Duplex resonator 14
Dynamic frequency-temperature effect 21, 38

Eigenvalue 24, 31, 169
Eigenvector 169
Elastic compliance 195, 202
Elastic constants 5, 194
Electric displacement 23, 180
Electric field 23, 180
Electric polarization 180
  per unit mass 183
  per unit volume 180
Electric potential 23, 180
Electric susceptibility 194-5
Electrodes
  air-gap 78
  base-plated 78
  losses due to 32, 113
  mass loading 56, 81
  materials 79
Electromechanical coupling factor 27, 36, 40
Energy trapping 45, 55
Equivalent circuit
  crystal 30, 91-3
  coupled resonators 152-4
  element values 30-2, 91-2
Etching 77
Evanescent waves 48, 212
Evaporation, thin film electrodes 78-9
Extensional mode resonators 9, 11
Extensional (E) waves 46

Face shear resonators 12, 19
Face shear (FS) waves, 46, 211
FC-cut 37-8
Field equations 23, 188, 192
Figure of merit 96
Filter types 143
Filter design
  image parameter method 144
  synthesis by 'exact' methods 144
Flexural mode resonators 14
Flexural (F) waves 46, 217
Frequency
  anti-resonance 28, 92
  cut-off 48, 55, 212
  fractional frequency difference 29-30, 34
  parallel resonance 92
  series resonance 28, 92
  tolerances 83
Frequency changes
  with drive level 119

INDEX 227

with stress 38-9, 119
with temperature 9-11, 13, 18-9, 21, 32, 37, 42, 114
Frequency constant 11, 37
Frequency stability
  in oscillators 131, 134-6
  long term 77, 117-8
  short term 118, 135-6
Frequency-thickness constant 10, 13, 36, 40
Fundamental mode 10

Gauss' theorem 168
Goniometer, X-ray 66
Group delay 141
Group velocity 46
Growth, cultured quartz 3
GT-cut 15
  original form 15
  miniature form 18

Headers, crystal 81
High-quartz 3
Holders, crystal 107
  ceramic 80, 85
  glass 80, 85
  metal, cold weld 80, 85
  metal, solder seal 80, 85
  metal, resistance weld 80, 85
Holder design 80

Impedance
  crystal 93-4
  motional 96
  at resonance, anti-resonance 103
Inductance, motional 30
Inflection temperature 115
Intermediate band filters 144
Intermodulation 119, 158
Inversion temperature 3
IT-cut 37

Lapping 70
  abrasives 74
Left quartz 4, 5
LC-cut 37
Load capacitor 93, 103
Long term frequency stability 77, 117-8, 135-6
Loop gain 127
Loop phase 127
Low-quartz 3
Length extensional modes 11

Mass loading 56, 81, 118
Material constants 5, 194
  density 3, 5, 7
  dielectric constants 5-7, 195
  elastic constants 5-7, 194

piezoelectric constants 5-7, 194
  temperature coefficients 6-8
  thermal expansion coefficients 6-7
Matrix notation 202
Maximum admittance, frequency of 101
Maxwell stress tensor 184
Melting point 3
Microwave filters 143
Miller indices 65
Miller oscillator 130
Mode coupling 9, 12, 14-5, 19, 51
Modes of vibration 8
  a, b, c thickness 37
  extensional 9, 11
  face shear 12, 19
  flexural 14
  fundamental 10
  length extensional 11
  overtone 9
  thickness extensional 9
  thickness shear 13, 19
Monolithic filter 143, 151
Motional parameters
  capacitance C_1 30, 91, 110
  inductance L_1 30
  resistance R_1 32, 113
Mounting 79-81
Mounting losses 113
MT-cut 22

Narrowband filters 144
Natural modes 92
Natural quartz 3-5
Network analyser 123
Noise: cf. Frequency stability, short term
Non-linear effects 119, 135, 158
NT-cut 22

Optical activity 4
Optical processing 63
Optical twinning 5
Orientation
  of bars 66-7
  of blanks 66-7
  of plates, specification of 35
Oscillators 127
  Clapp 130
  Colpitts 130
  conditions for oscillation 127, 129
  feedback 127
  Miller 130
  oven controlled 139
  Pierce 130
  temperature compensated 137-8
  temperature controlled 139
  voltage controlled

Parallel resonance 92
Parametral plane 65

Phase characteristic, crystal resonator 99
Phase noise: cf. Frequency stability, short term
Phase offset method, measurement of motional parameters 122-3
Phase zero measurement system 121
Photolithographic techniques 18, 86, 87
Piezoelectric constants 5-7, 194-5
Piezoelectricity
  explanation of effect 160
  general theory 179
  linear theory 190
Pin lap 72
Pi-network 121-2
Planetary lap 71
Plateback 81
Polarization 180
Polishing
  mechanical 70, 74
  chemical 77
Polylithic filter 143, 151
Potential, electric 23, 180
Pulling sensitivity 31, 92, 95

Q factor
  crystal unit 32, 91, 96, 112
  minimum requirement for narrowband filter 156
  and oscillator stability 134
Quartz
  alpha 3-5
  beta 3, 5
  cultured 3, 5
  high 3
  left 4, 5
  low 3
  natural 3-5
  occurrence 3
  properties 3
  right 4-5
  twinning 5

Radio communications 143
Rational indices, law of 65
Rayleigh surface wave 218
Reflection, X-ray 64
Resistance, motional 32, 113
Resonance
  electrical 28, 101
  mechanical 9, 11, 13, 27, 49
Right quartz 4-5
Rotated Y-cuts 18, 21, 39
Rounding 69
RT-cut 37

Sawing 69
SAW filters 143
SC-cut 21, 37-8
Sealing 85

Second level of drive 119
Seed crystal 3, 69
Series resonance 28, 92
Shunt capacitance 30, 110
Sign convention 5, 37
Sleepy crystal 119
Solubility, quartz 3
Stability: cf. Frequency stability
Standing waves 9
Static capacitance 30, 110
Strain
  elastic 8-9, 173
  thermally induced 6, 33
Strain tensor
  finite 173
  infinitesimal 23, 191
Stress, elastic 176
Stress tensor 178
Substitution, measurement by 120
Surface finish 70, 118-9
  effect on ageing 77
  effect on Q 32, 77, 113
Symmetric modes 26

Temperature coefficients
  material constants 6-7, 32
  frequency 9-11, 13, 18-9, 21, 32, 37, 42, 114
Temperature compensated oscillators 137-8
Temperature controlled oscillators 139
Temperature gradients 21
Tensors 163
Time constant 32, 112
Thermal expansion 6-7, 32
Thermally induced strain 6, 33
Thickness extensional (TE) modes 9, 37
Thickness extensional (TE) waves 46
Thickness modes 9, 18-9, 23, 37, 210
Thickness shear (TS) modes 12-3, 37
Thickness shear (TS) waves 46, 217
Thickness twist (TT) waves 46, 211
Transformation of coordinates 162, 197
Transition temperature 3
Transmission measurements 120-1
Trapped energy resonators 45, 55
Tuning forks 14
Turning points 115
Turnover temperature 115
Twinning
  electrical 5
  optical 5

Unwanted modes
  in crystals 59, 117
  in filters 157
UV-ozone cleaning 78

Vacuum deposition 78-9

INDEX 229

Varactor diode 138, 140
Vector voltmeter 121
Velocity
  group 46
  particle 174
  phase 25, 28, 46
  wave 10, 46
Vibration, modes of 8
Voltage controlled oscillator 140

Wafer processing 69
Waves
  evanescent 48, 212
  extensional (E) 46
  face shear (FS) 46, 211
  flexural (F) 46, 217
  longitudinal 9, 11, 25
  plane 23
  plate 45, 208
  shear 13, 25
  thickness extensional (TE) 46
  thickness shear (TS) 46, 217
  thickness twist (TT) 46, 211
  transverse 13, 25
Wideband filters 144

X-cut resonator family 9, 11-2, 22
X-rays
  diffraction 64
  goniometer 66
  orientation 64
  reflection 64
  transfer jigs 69

Y-cut 12-3

Zero phase measurement method 121
Zero phase pi-network 121-2